£14

TOJ

(J.

GW00724521

X

*Finite Elements in
Electrical and Magnetic
Field Problems*

Finite Elements in Electrical and Magnetic Field Problems

1980

Edited by

M. V. K. Chari
Corporate Research & Development
General Electric Company,
Schenectady, New York
and

P. P. Silvester
Department of Electrical Engineering,
McGill University, Montreal

A Wiley – Interscience Publication

JOHN WILEY & SONS
Chichester · New York · Brisbane · Toronto

Library of Congress Cataloging in Publication Data:
Main entry under title:

Finite elements in electrical and magnetic field problems.

  (Wiley series in numerical methods in engineering)
  'A Wiley–Interscience publication.'
  Includes index.
  1. Electric engineering—Mathematics.   2. Finite element method.
3. Electric fields.   4. Magentic fields.   I. Chari, M. V. K.   II. Sil-
vester, Peet.
TK153.F48   621.3   79-1037
ISBN 0 471 27578 6

Photoset and printed in Malta by Interprint Limited

Contributory Authors

Dr. M. V. K. Chari — General Electric Company,
1 River Road, Schenectady,
New York 12345, U.S.A.

Dr. Z. J. Csendes — General Electric Company,
1 River Road, Schenectady
New York 12345, U.S.A.

Professor
A. L. Frisiani — Ist. Elettrotecnica,
Viale Causa 13,
16 145 Genoa, Italy.

Professor
R. H. Gallagher — College of Engineering,
The University of Arizona,
Tucson, Arizona 85721, U.S.A.

Professor
P. J. Lawrenson — Department of Electrical Engineering,
University of Leeds,
Leeds LS2 9JT, U.K.

Dr. B. H. McDonald — Atomic Energy of Canada Ltd,
Pinawa, Manitoba, Canada.

Dr. T. J. E. Miller — General Electric Company
1 River Road, Schenectady,
New York 12345, U.S.A.

Dr. G. Molinari — Department of Electrical Engineering
University of Genoa,
Viale Causa 13
16 145 Genoa, Italy.

Dr. J. C. Sabonnadière — Inst. Nat. Polytechnique,
B. P. 15 Centre de Tri,
38040 Grenoble-Cedex, France,

Dr. P. P. Silvester — Department of Electrical Engineering,
McGill University,
Montreal H3A 2A7, Canada.

Dr. C. W. Trowbridge — Building R25, Rutherford Laboratory,
Chilton, Didcot, 0X11 0QX, U.K.

v

DR. A. VIVIANI

*Department of Electrical Engineering,
University of Genoa,
Viale Causa 13,
16 145 Genoa, Italy.*

PROFESSOR
A. WEXLER

*Department of Electrical Engineering,
University of Manitoba, Winnipeg,
Manitoba R3T 2N2, Canada.*

PROFESSOR
O. C. ZIENKIEWICZ

*Department of Civil Engineering
and Dean of Applied Science,
University College of Swansea, U.K.*

Contents

Contents

Editors' Preface

Ever since the early days of development of the theory of electricity and magnetism, accurate solution of initial and boundary value problems in electrical engineering has been a major objective. Until the sixties, efforts were mainly directed towards obtaining closed form analytical and analog solutions based on simplified modelling of the boundary value problems and the associated differential equations.

With the advent of digital computers, numerical solutions such as finite difference schemes, finite elements, and integral equations have gained currency. However, the development and application of these techniques have been sporadic and generally problem oriented. Although nearly fifteen years have passed since the first efforts, many queries have remained unanswered and many more have been identified.

An International Conference on Numerical Methods in Electric and Magnetic Field Problems was held in Sta. Margherita, Italy (June 1–4, 1976) under the joint auspices of the International Center for Computer aided design of the University of Genoa and the International Journal for Numerical Methods in Engineering. The invited papers given at that conference, together with three additional chapters, form the basis of this book.

Within the limits of steady state finite element and integral equation solutions and linear time varying solutions of electric and magnetic field problems, the reader is here presented with a broad picture of current thought and research in this area.

The chapters are arranged into three principal categories, namely introductory and concept development; applications and advanced techniques; and specific methods. These are described in some detail in the following.

The introductory chapter reviews a number of practical situations arising in electrical engineering which can be formulated as initial and boundary value problems. Different methods of solving the associated partial differential equations by analytical, analog, and numerical methods are surveyed. These range from the classical separation of variables technique and its variants as applied to transformers and inductors by Roth and Rogowski, conformal mapping techniques, conducting paper plots, and finally computer based

ix

numerical methods such as finite difference schemes and finite elements. A brief discussion of the relative merits of the respective methods is presented.

In Chapter 1, the concept of the finite element method and its relative merits vis-à-vis other established methods such as finite differences and boundary integral methods are discussed. Different finite element approximations arising from variational principles, weighted integral expressions, Lagrangian multipliers and penalty functions, virtual work principles, and others are surveyed. General principles underlying the finite element approximation for two and three dimensional electric and magnetic field analysis are described. An application of the method for analysing transformer magnetic fields using a scalar potential approximation is presented.

The electromagnetic field is basically a tensor quantity, which may be described in a variety of ways. In Chapter 2, criteria for choosing the field representations in finite element analysis are discussed. First, any field representable by a linear combination of the finite element basis functions should be physically realizable. Secondly, the describing equations should lend themselves to a variational or projective solution which will include natural boundary conditions occurring at source-free surfaces or material interfaces. Unfortunately, no single fully satisfactory description of the electromagnetic field is known. Skilful choice of field representations, different for every new class of problems, is thus essential, so as to achieve simplifications and computational savings. Some of the more usual representation of fields are reviewed.

The finite element method involves the subdivision of the field region into subdomains or finite elements and approximation of the field in each element in terms of a limited number of parameters. Polynomial expansions are the customary choice. In Chaper 3, the different methods of selecting shape functions for elements are reviewed and the criteria for their choice are stated. The properties of polynomial functions are discussed as relevant to finite element approximations. The concept of isoparametric mapping which enables the construction of elements with curved or distorted boundaries is described.

Chapter 4 describes the various aspects of software engineering required in finite element analysis. In a typical program package, the mathematical software occupies only a small portion, while most of the code is devoted to problem definition and data handling; control of programme flow sequence and error checking; and post-processing the solutions into forms useful for engineering purposes. Software engineering seeks to ensure that programme packages serve their intended mathematical purpose while communicating with the user in a fairly problem-oriented manner. It strives to maintain reasonable programme portability and flexibility consistent with economical use of the computing hardware systems for which the programmes are intended. This chapter reviews the current trends in both hardware and software design and suggests that future software packages should be designed with a high degree of modularity and standardization of file structure.

In Chapter 5, the development of the finite element method for solving two and three dimensional electromagnetic fields in electric machines and devices is presented. Various applications of the technique for linear and nonlinear problems are discussed. Some of the areas surveyed are magnetic fields in electrical machinery cross-sections and the end-region; transformers; diffusion problems and eddy-current analysis in conducting media and electrostatic applications.

Eddy currents, which are often viewed as only a harmful phenomenon, have their uses in industrial applications such as in induction heating, magnetic propulsion and suspension, and others. In Chapter 6, different analytical methods for accurately predicting eddy currents in various practical situations are reviewed. These include series solutions for finite regions (eigenvalue problems) and infinitely extending structures (Fourier analysis). Analysis of three-dimensional problems by orthogonal function methods is also discussed. For nonlinear problems, the Galerkin projection method is recommended. It is concluded that for the general eddy current problem, no single all-comprehensive technique exists other than a careful analysis of each individual problem.

Chapter 7 summarizes the application of the high-order polynomial finite element method to electromagnetic field calculations. It provides a basic review of the development of the method indicating the motivation for its construction and outlines its algebraic development. Problems encountered in the computational implementation of the method are described and a bibliography of the published applications of the high-order polynomial finite element method in electromagnetics is provided.

Chapter 8 shows how the Fast Fourier Transform technique can be used with advantage to solve transient electromagnetic diffusion fields. Attention is focussed on the considerations underlying the application of the method to solve practical engineering problems. The method is effective because the space and time solutions to the field problem can be separately handled. This technique requires evaluation of a single frequency-response function, which is then used repeatedly with the FFT for each time function, yielding utmost computational economy.

There are many problems encountered in practice which do not clearly lend themselves to formulation in either integral or differential equations. Chapter 9 indicates one possible avenue of approach in such cases: part of the problem is dealt with in integral, part in differential form. Requiring the partial solutions to match, imposes mutual constraints on the two systems of equations, which are usually best solved by variational techniques.

In Chapter 10, various integral equation methods are described. For magnetostatic problems, three formulations are considered in detail: (a) the direct solution method for the magnetic field distribution in permeable materials, (b) a method based on a scalar potential, and (c) the use of an integral equation derived from Green's theorem, i.e. the so-called Boundary Integral Method (BIM). In case (a), results are presented for both two- and three-dimensional nonlinear problems and comparisons are made with measurements. Methods (b) and (c) lead to a more economical use of the computer than (a); for these, preliminary results for simple cases are included. Techniques for solving the eddy current problem are discussed, and computed results based on a vector potential formulation are presented.

The finite element art has had a marked impact on electromagnetic field analysis in the past decade, and will no doubt continue to do so. While most of the early work dealt with scalar, two-dimensional, static fields, the chapters presented here clearly point the way to broader problems. No doubt many new methods and many new problems will appear in the next decade. The trend to increased use of finite element methods will surely continue, fuelled by a rapidly broadening range of available computing resources, and motivated by increased acquaintance with their power—*avec le manger bient l'appetit.*

The editors wish to express their appreciation for the opportunity the authors have granted them to engage in this most rewarding in-breadth study of the finite element field. They wish also to thank the editorial staff of John Wiley & Sons Limited for their extensive counsel and assistance.

Schenectady and Montreal   M. V. K. CHARI
25 March 1979   P. P. SILVESTER

Introduction

A. L. Frisiani, G. Molinari, and A. Viviani

Boundary-value problems of mathematical physics occur in practically every engineering application. Different aspects are present, for instance, in structural analysis, heat transfer, fluid flow, electromagnetic fields, and they are indeed of great interest in all practical design problems.

In the design process, even in its roughest phases, the designer tries to define, by successive hypotheses and approximations, suitable boundary-value problems and to find acceptably accurate solutions to them.

In the past, and still today in design problems involving only minor economic and technological difficulties, it was generally assumed that no interaction took place among the various fields, for instance between the electromagnetic field and the thermal one. Single-field problems were solved by means of approximate procedures, generally requiring considerable simplifications of geometries and materials involved. This procedure, which largely makes use of solutions previously determined, such as uniform field solutions, originated the lumped parameter approach, so widely employed especially in electrical engineering.

The development of advanced technologies and the increase in dimensions and costs of many engineering systems have made necessary a parallel development of more general and accurate computation techniques. It has become increasingly important to obtain a deeper knowledge of the spatial distribution of vector and tensor fields, either for improved accuracy in evaluating integral parameters, or to determine and localize maximum values which generally denote critical stress conditions in materials.

In the last decades, an impressive growth of the rating of electrical systems, and consequently of their dimensions and costs, has taken place. About fifteen years ago, the maximum rating of turbo-alternators was in the range of 200 MVA, while machines in the range of 1 to 1.5 GVA have recently been built, and even larger ones are under consideration. Similar increases have taken place in the rating of transformers, cables, and many other electrical devices.

This situation poses new problems to the designer. For instance, in the past, valuable help was forthcoming from experimental data obtained from working systems. However, the increase in ratings and related costs has greatly reduced the availability of experimental data.

On the other hand, it is impractical to increase the volume of electrical devices in proportion to their rating, for obvious economic reasons. It is, therefore, necessary to increase stresses in materials, and to adapt new design criteria to the changed requirements. As a consequence, new design solutions become possible; for instance, the use of superconductors has been proposed in the case of electrical machines and cables. Frequently, new devices are also notably different in shape from previous ones; consequently, old computation procedures become inapplicable.

Therefore, an increase in accuracy of theoretical performance prediction is necessary, especially in view of the complexity of the geometry of problems involved and the characteristics of materials used.

For instance, a better knowledge of space and time distributions of both electric current and magnetic flux in an electric machine is necessary in order to obtain a reliable description of power losses and electromechanical stresses, from which the thermal and mechanical design of the machine must be derived. But this may require a generalization of steps of the computation procedures, such as taking into account anisotropy, saturation, and hysteresis of magnetic materials, the laminar structure of magnetic cores, the presence of slots and air gaps, the spatial distribution of conductors and dielectric materials, and the influence of frame materials, which can no longer be regarded as electrically and magnetically passive. Besides, the computation must be performed under time-varying conditions, which are generally non-sinusoidal on account of the nonlinearities of materials or due to the transient conditions of applied voltages or torques. Furthermore, the general hypothesis according to which the whole machine is isothermal can no longer be accepted. This hypothesis allows us to treat the electric and the magnetic fields independently, by using Maxwell's equations and constitutive relations in a form independent of temperature. If the hypothesis is no longer valid, constitutive relations must also contain temperature, and thermal field equations must be added to Maxwell's equations. Similar problems arise when determining stresses in dielectric materials.

The above considerations are applicable to high power traditional machines, such as transformers, turbo-generators, or salient pole alternators. They also apply equally well to nontraditional versions of the machines, as well as to other electrical apparatus, such as high-voltage equipment, high-field magnets, direct-current generators, and motors with limit performances. They can be extended to the implementation of interesting nontraditional electrical devices, such as linear motors, levitation systems and, possibly, special magnets for nuclear fusion, MHD generators, and energy storage. In these cases, we have to deal with problems in which geometries involved are generally not suitable for traditional approaches, and, therefore, require powerful calculation procedures.

Another electrical engineering area requiring efficient computation techniques is electronics, especially in the high-frequency range. The technologies of semiconductor devices and integrated circuits, which have produced the rapid growth of the electronics industry, have also made necessary specific requirements for solving problems with general forms of nonlinearities in semiconductor transport equations, and for taking into account, as far as possible, two- or three-dimensional effects. Other types of electronic device (for instance, in the area of wave propagation systems, or in the presence of plasmas or of special materials, such as piezo-electric ones) also require new computational techniques.

The conditions outlined above have strongly influenced the historical development of computational methods for boundary-value problems, and explain the present interest and activities in such a field, particularly in numerical procedures.

The methods used from Maxwell onwards for the solution of boundary-value problems can be divided into four categories: analogue, graphical, analytical, and numerical.

Analogue procedures consist in obtaining the unknown field by experimental measurements on an analogue of the field region, i.e. on a field region governed by the same equations and with the same boundary and interface conditions. These procedures have generally been used only for Laplace's equation under two- or three-dimensional conditions. In fact, it has been practically impossible to model inhomogeneities, nonlinearities, and so on, by means of media different from the ones involved in the real problem. Besides, in its three-dimensional version (such as, for instance, electrolytic tank or resistance network), this method is rather expensive and cumbersome, whereas the more convenient versions (such as, for example, graphitic paper or elastic membrane) are restricted to two-dimensional fields.

Graphical procedures have long been used, but they are restricted to Laplace's equation for two-dimensional geometries because they are generally based on the properties of analytic functions (see, for instance, the Lehmann method). It should be added that their accuracy is limited even when they are carefully applied.

The development of analytical methods advanced a good deal while numerical methods were still in their infancy. These methods are still widely used and include series solution and conformal mapping techniques. Other methods in vogue are integral equations, variational formulations, or approaches specific to various problems. The last mentioned ones, such as the method of images or the inversion method, are generally applicable to simple geometries and materials. In such cases, solutions are found by inspection and are based on known solutions to analogous problems, or by the use of symmetry conditions, etc.

Series solutions are generally obtained by the so-called method of separation of variables. This method can be mainly applied to Laplace or Helmholz equations in two- or three-dimensional problems, and can also be applied in time-dependent problems, for instance in the ones governed by diffusion or wave equations. Even if such a method can be considered as being more general than the previous ones, it suffers from severe restrictions essentially related to the treatment of boundary and interface conditions. In practice, its use requires the existence of a suitable coordinate system, which must fulfil two conditions: (i) every boundary or interface surface must coincide with an equi-coordinate surface; (ii) the coordinate system must allow the separation of variables. Both these conditions are rarely satisfied in complex problems. The calculations require the use of special functions (such as Bessel, Legendre, elliptic) which often are not easily handled.

The separation of variables method can also be applied to inhomogeneous equations like the Poisson equation. In this case. it is necessary to add a particular integral of the inhomogeneous equation to the general integral of the corresponding homogeneous one: the particular integral can generally be computed by volume or surface integration. However, it is often quite difficult to perform such an integration analytically, and special techniques have been introduced to solve problems, even for very simple geometries. We recall here Rogowski's and Roth's methods, developed for the solution of magnetic fields in transformers and inductors.

The solution of field problems by conformal mapping is another analytical approach which has been extensively used. It is based on the properties of analytic complex functions, so that it can only be applied to problems which can be reduced to Laplace's equation in a two-dimensional region. In such a domain, conformal mapping may often be more powerful than the series method because it can yield closed-form solutions for more complicated regions. Severe limitations have, however, to be placed on the geometry of problems to avoid difficulties in the integration of complex functions. Analytic functions can also be used to generate coordinate transformations preliminary to the handling of equations by other methods, but this application is also limited.

The aforesaid is a brief description of analytical methods. This category is very broad and includes algorithms that generalize and extend the above procedures, even if laboriously. The major deficiency of the analytical methods is the lack of generality. The classes of problems for which analytical solutions exist that can be considered somewhat general are by far the simplest ones and many algorithms are applicable only to two-dimensional and steady-state problems. Besides, algorithms for inhomogeneous and nonlinear problems are practically non-existent excepting for some extremely simple and special problems. Another notable deficiency of analytical methods lies in the effort required in obtaining the field solutions, such as developing special algorithms and discovering artifices.

The deficiencies of analytical methods are to a large extent eliminated in numerical methods, which have come to the fore with the advent of large digital computers. The principal numerical methods that are in vogue can be subdivided into finite difference schemes, image methods, integral equation techniques, and variational formulations. ~ finite element method.

Finite difference schemes were the first to be widely used, in practice since about 1940, even if they have been traced back to Gauss.[1] For one-dimensional problems a complete finite-difference approach can already be found in the graphical string polygon method for finding deflections of beams developed by Mohr in 1868.[2] The first application to two-dimensional problems was made by Runge in 1908.[3] There was a parallel development of methods of solution for the large algebraic equations resulting from finite-difference schemes; it was based on the fundamental contributions by Gauss,[4] Jacobi,[5] and Seidel,[6] and on the results obtained by Richardson[7] and Liebmann[8] in the area of iterative methods, and on the works by Gauss,[9] Doolittle,[10] and Choleski[11] in the area of direct methods.

As we can see, numerical methods had been defined long before their wide use, which began with the advent of high-speed computers.

The basis of finite-difference schemes is the replacement of a continuous domain with a grid of discrete points ('nodes'), the only ones at which the value of unknown quantities are computed. The reduction of the equations and of the boundary or interface conditions, defined in the continuous domain, to the discretized equations valid for the nodes is performed by means of various algorithms, which replace derivatives and integrals with 'divided-difference' approximations obtained as functions of the nodal values. This can be accomplished, for instance, by using interpolating functions, which are not defined in specific subdomains but simply in the neighbourhood of a node.

In its traditional versions, the grid is a regular one: that is, a rectangular grid with nodes at the intersections of orthogonal straight lines or a polar grid with nodes at the intersections of orthogonal circles and radii. This restrictive approach, which simplifies the discretization algorithms, is not necessary. However, the use of general curvilinear grids (or of irregular ones, seldom proposed in the past) has not been successful, so that regular grids are the only ones in practice to date. As a consequence, severe difficulties are encountered in solving many problems using finite-difference schemes, and therefore their efficiency is considerably limited. This is essentially due, apart from other less important problems, to geometrical reasons related to the fitting of the grid to the shapes of boundaries and interfaces involved. In fact, a regular grid is not suitable for problems with very steep variations of fields. The grid must indeed be denser in regions of high field gradients, and this requires either a very large number of nodes, so that computation time and memory requirements are significantly increased, or a complex algorithm to increase the density of the grid artfully. Moreover, a regular grid is not suitable for curved boundaries or interfaces, because they intersect gridlines obliquely at points other than nodes. This may not be a problem under Dirichlet boundary conditions, but poses difficulties under Neumann boundary conditions or under interface conditions involving normal derivatives. These situations require sophisticated interpolation schemes which are difficult to implement in an automatic form, and complicate the solution of algebraic equations resulting from the discretization.

In spite of these shortcomings, finite-difference schemes, in their traditional version, monopolized the area of numerical methods in electrical engineering practically up to 1970. In any case, they led to valuable results: we may recall here the work of Erdélyi's group at the University of Colorado,[12] and the activities of Fritz, Müller, and Wolff in the area of alternators and d.c. machines.[13]

Alternative methods have also been proposed in the area of image methods and integral equation techniques which have intensive but limited application in special areas. We mention here particularly, in the area of image methods, the work of Prinz and Singer's group.[14] They developed the charge method, in which the values of image charges are computed to simulate three-dimensional high voltage fields. Likewise, in the area of integral equations, we recall the activities of Trowbridge's group at the Rutherford Laboratory in the computation of three-dimensional structures of magnets.[15] Both methods seem mainly oriented towards three-dimensional nonsymmetrical problems, and, in this connection, they could provide interesting developments.

In other engineering areas, particularly in civil engineering, the drawbacks of traditional finite-difference schemes have been recognized at an early stage and alternative methods have been developed, such as variational procedures. This process, which can also be traced back to the past century, has led to the modern form of the finite-element method, and can be considered to have been completely established in the late fifties. Actually, the term 'finite-element method' was first used in a paper by Clough in 1960.[16]

Variational methods consist in formulating the equations of boundary-value problems in terms of variational expressions called 'energy functionals', which, in electrical applications, often coincide with the energy stored in the field. The Euler equation of this functional will generally coincide with the original partial differential equation. In the finite-element method, the field region is subdivided into elements, that is, into subregions where the unknown quantities, such as, for instance, a scalar or a vector potential, are represented by suitable interpolation functions that contain, as unknowns, the values of the potential at the respective nodes of each element. The minimization of the energy functional by the use of such interpolation functions generates an algebraic system of equations, as in the finite-difference methods, and the potential values at the nodes can be determined by direct or iterative methods.

The first application of the finite-element method in electrical engineering was made by Winslow in the analysis of saturation effects in accelerator magnets,[17] and the first general nonlinear variational formulation of electromagnetic field problems in electrical machines is due to Silvester and Chari.[18] Afterwards, the method spread rapidly, and now it can be considered to be the fastest growing technique in the area of electrical engineering.

The finite-element method has since been applied to various electrical problems, and we list here some of them. Transformers were studied by Silvester and Chari[18] and Andersen.[19] Applications to turbo-alternators were made by Chari and Silvester,[20] Brandl, Reichert, and Vogt;[21] end-region fields of such machines were studied by Howe and Hammond,[22] Chari, Sharma, and Kudlacik,[23] and Okuda, Kawamura, and Nishi;[24] and salient pole generators were studied by Glowinski and Marrocco.[25] Eddy-currents were treated by Silvester and Haslam[26] in magneto-telluric problems; by Chari[27] in magnetic structures; by Foggia, Sabonnadière, and Silvester[28] in linear induction motors. Anisotropic magnetic problems were treated by Wexler[29] in a reluctance machine. Electrostatic problems were especially studied by Andersen.[19]

After such contributions, which evidenced its merits, the finite-element method will certainly become more widespread in the near future. Flexibility is its greatest advantage with respect to traditional finite-difference methods. Elements can have various shapes, and can be easily adapted to any shape of boundary and interface geometries. Flexibility can even be a little excessive, since it introduces a strong influence of the user on the results obtained, so that it must be exploited with caution. Besides flexibility, another advantage lies in the form of the algebraic system of equations obtained, which generally has a symmetric positive definite matrix of coefficients. This property is not ensured in finite-difference schemes, and can facilitate the solution. Another advantage is claimed to be simpler programming on account of the easier introduction of boundary conditions.

Apart from these remarks, the comparison among different methods is always a delicate matter. This is particularly true for finite-difference and finite-element schemes, the comparison of which led, in recent years, to a lively dispute among their supporters with arguments which often turned out to be groundless. Indeed, the merits of any method depend essentially on its particular implementation, and it often happened that comparison was made between a refined implementation of the method supported and a rough one of the method rejected. Several comparison criteria have been proposed, but very few can be considered as valid. For instance, some workers have used the same nodes in both methods; this criterion, however, perverts the nature of the finite-element method whose main feature is the use of irregular grids. Other workers used grids with an equal number of nodes, or with a number of elements in the finite-element method equal to the number of domains in the finite-difference one. Still others based comparison on conditions that do not depend on the discretization method used, but on the procedure for solving the algebraic system of equations.

All the aforesaid criteria, and others not mentioned here, appear to be irrelevant, because they give pre-eminence to data which are not important to the user. Generally, the user is only interested in a problem and its solution, which must be obtained with a given precision and at the lowest possible cost. Comparison of numerical methods must, therefore, be based on the criterion of equal precision. This precision, in turm, can be evaluated as the maximum error of the unknown potential, or as the error of integral quantities to be computed. The maximum error should be calculated taking into account the extrapolation procedures typical of the method used. Maximum or suitably defined mean errors at the nodes are not reliable and should not be used. Of course, comparison should be made with theoretical results, if any, or with experimental data; however, experimental data can introduce other errors which are not due to the numerical method but to an unsuitable modelling of physical reality, and consequently can originate less certain results. Once a precision criterion has been decided upon, the methods must be evaluated mainly on the basis of costs. Amongst other parameters of varying importance, we mention computer memory requirements and ease of programming (if no general programme is available).

Following such criteria, conclusions could generally be drawn for more or less extended classes of problems. It is practically impossible to find a method that is superior to others for every type of problem.

At present, the salient features outlined above of both the traditional finite-difference and finite-element schemes seem to indicate that the finite-element method is more powerful for a large class of problems with complicated geometries, whereas for simpler geometries, such as domains made up of rectangles in a suitable coordinate system, the two methods are similar, and preference can be based on particular conditions, depending on the type of boundary and interface conditions, on materials, and so on.

Taking into account that finite-element methods are relatively 'young' in the electrical engineering area, significant developments can be expected for them in the future. On the other hand, finite-difference methods are older but have never been exploited completely and are at present experiencing a revival with the use of irregular grids and curvilinear gridlines. Therefore, no conclusion can be considered final; a fascinating field of activity is open, promising interesting developments in the future.

REFERENCES

1. Gauss, C. F. (1823). In '*Carl Friedrich Gauss Werke*', Vol. 9, Göttingen, 1873, 278–281.
2. Mohr, O. (1868). 'Beitrag zur Theorie der Holz und Eisenkonstruktion', *Z. Arch.-u. Ing.-Ver.*, Hannover.

3. Runge, C. (1908). 'Uber eine Methode die partielle Differentialgleichung Δu =constants numerisch zu integrieren', *Z. Math. u. Phys.*, **56**, 225–232.

4. Gauss, C. F. ($\sim$ 1825). In C. L. Gerling, *'Die Ausgleichs-Rechnungen der practischen Geometrie'*, Hamburg and Gotha, 1843.

5. Jacobi, C. G. J. (1844). In C. G. J. Jacobi, *'Gessamelte Werke'*, Vol. 3, Berlin, 1884, 467–478.

6. Seidel, L, (1874). 'Uber ein Verfahren die Gleichungen auf welche die Methode der Kleinsten Quadrate führt, so wie lineäre Gleichungen überhaupt durch successive Annäherung aufzulösen', *Abhandl. bayer, Akad. Wiss.*, **11**, 81–108.

7. Richardson, L. F. (1911). 'The approximate arithmetical solution by finite differences of physical problems involving differential equations, with an application to the stresses in a masonry dam', *Phil. Trans.*, **210A**, 307–357.

8. Liebmann, H. (1918). 'Die angenäherte Ermittelung harmonischer Funktionen und konformer Abbildungen', *S. B. Bayer. Akad. Wiss., Math. Phys. Klasse*, 385–416.

9. Gauss, C. F. (1826). In *'Carl Friedrich Gauss Werke'*, Vol. 4, Göttingen, 1873, 55–93.

10. Doolittle, M. H. (1878). 'Method employed in the solution of normal equations and the adjustment of a triangulation', *U.S. Coast Geodetic Survey Rept.*, 115–120.

11. Choleski, A. L. (1916). In Benoit, 'Note sur un methode, etc. (Procédé du Commandant Cholesky', *Bull. geodésique*, 1924, 67–77.

12. Fuchs, E. F. and Erdélyi, E. A. (1973). 'Nonlinear theory of turboalternators. Parts I, II', *IEEE Trans. on PAS*, **PAS-92**, 583–599.

13. Müller, W. and Wolff, W. (1973). 'Numerische Berechnung dreidimensionaler Magnetfelder für grosse Turbogeneratoren bei feldabhängiger Permeabilität und beliebiger Stromdichteverteilung', *ETZ-A*, **94**, 276–282.

14. Singer, H. Steinbigler, H., and Weiss, P. (1974). 'A charge simulation method for the calculation of high voltage fields', *IEEE Trans. on PAS*, **PAS-93**, 1660–1668.

15. Simkin, J. and Trowbridge, C. W. (1976). 'Application of integral equation methods for the numerical solution of magnetostatic and eddy current problems,' *Rutherford Laboratory, Report RL-76-041*.

16. Clough, R. W. (1960). 'The finite element method in plane stress analysis', *Proc. 2nd Conf. Electronic Computation, ASCE, Pittsburg, Pa.*

17. Winslow, A. M. (1965). 'Magnetic field calculation in an irregular triangular mesh', *Lawrence Radiation Laboratory (Livermore, California), UCRL-7784-T*, Rev. 1.

18. Silvester, P. and Chari, M. V. K. (1970). 'Finite element solution of saturable magnetic field problems', *IEEE Trans. on PAS*, **PAS-89**, 1642–1651.

19. Andersen, O. W. (1973). 'Laplacian electrostatic field calculations by finite elements with automatic grid generation', *IEEE Trans.*, **PAS-92**, No. 5.

20. Chari, M. V. K. and Silvester, P. (1971). Analysis of turbo-alternator magnetic fields by finite elements', *IEEE Trans.*, **PAS-90**, 454–464.

21. Brandl, P., Reichert, K. and Vogt, W. (1975). 'Simulation of Turbogenerators on Steady State Load', *Brown Boveri Review*, **9**.

22. Howe, D. and Hammond, P' (1974). 'The distribution of axial flux on the stator surface of the ends of turbogenerators', *Proc. IEE*, **121**, No. 9, 980–990.

23. Chari, M. V. K., Sharma, D. K., and Kudlacik, H. W. (1976). 'No load magnetic field analysis in the end region of a turbine generator by the method of finite elements', *IEEE Winter Meeting, New York, Paper No. A 76 230–3*.

24. Okuda, H., Kawamura, T., and Nishi, M. (1976). 'Finite-element solution of magnetic field and eddy current problems in the end zone of turbine generators', *IEEE Winter Meeting, New York, Paper No. A 76 141–2*.

25. Glowinski, R. and Marrocco, A. (1974). 'Analyse numerique du champ magnetique d'un alternateur par elements finis et surrelaxation ponctuelle non lineaire', *Computer Methods in Applied Mechanics and Engineering*, **3**, 55–85.

26. Silvester, P. and Haslam, C. R. S. (1972). 'Magnetotelluric modelling by the finite element method', *Geophysical Prospecting*, **20**, 872–891.
27. Chari, M. V. K. (1973). 'Finite-element solution of the eddy-current problem in magnet structures', *IEEE Trans.*, **PAS-93**, No. 1.
28. Foggia, A., Sabonnadiere, J. C., and Silvester, P. (1975). 'Finite-element solution of saturated travelling magnetic field problems', *IEEE Trans.*, **PAS-94**, No. 3.
29. Wexler, A. (1973). 'Finite-element analysis of inhomogeneous anisotropic reluctance machine rotor', *IEEE Trans.*, **PAS-92**, No. 1, 145–149.

Finite Elements in Electrical and Magnetic Field Problems
Edited by M. V. K. Chari and P. P. Silvester
© 1980, John Wiley & Sons Ltd.

Chapter 1

Finite Elements—The Basic Concepts and an Application to 3-D Magnetostatic Problems

O. C. Zienkiewicz

1.1 INTRODUCTION

What are finite elements? To supply an answer to this, certainly nontrivial, question we shall try to

(*a*) define the finite element method in its general form, and
(*b*) show its different facets which to a greater or lesser extent have been used to date.

In the field of structural and solid mechanics, the research developments of the last fifteen years (since the first mention of the 'finite element' name was made [1,2]) have become everyday practice in many stress analysis/structures situations.

1.2 THE FINITE ELEMENT CONCEPT

The finite element method is concerned with the solution of mathematical or physical problems which are usually defined in a continuous domain either by local differential equations or by equivalent global statements. To render the problem amenable to numerical treatment, the infinite degrees of freedom of the system are discretized or replaced by a finite number of unknown parameters, as indeed is the practice in other processes of approximation.

The original 'finite element' concept replaces the continuum by a number of subdomains (or elements) whose behaviour is modelled adequately by a limited number of degrees of freedom and which are assembled by processes well known in the analysis of discrete systems. Often at this early stage the model of the element behaviour was derived by a simple physical reasoning avoiding the mathematical statement of the problem. While one can well argue that such an approach is just as realistic as formal differential statements (which imply the possibility of an infinite subdivision of matter), we prefer to give here a more general definition embracing a wider sope.*

Ref.

Thus we define the <u>finite element process</u> as any approximation process in which :—

(a) the behaviour of the whole system is approximated to by a finite number, n, of parameters, a_j, $i = 1$ to n, for which

(b) the n equations governing the behaviour of the <u>whole system</u>

$$F_j(a_i) = 0 \qquad j = 1 \text{ to } n \tag{1.1}$$

can be <u>assembled</u> by the simple process of <u>addition</u> of terms contributed from all subdomains (or elements) which divide the system into physically identifiable entities (without overlap or exclusion). Thus

$$F_j = \Sigma F_j^e \tag{1.2}$$

where F_j^e is the element contribution to the quantity under consideration.

This broad definition allows us to include in the process both physical and mathematical approximations and, if the 'elements' of the system are simple and repeatable, to derive prescriptions for calculation of their contributions to the system equations which are generally valid. Further, as the <u>process</u> is precisely <u>analogous</u> to that used in <u>discrete system assembly</u>, computer programs and experience accumulated in dealing with discrete systems can be immediately transferred.

An important practical point of the approximation has been specifically excluded here. This concerns the fact that often <u>contributions of the elements</u> <u>are highly localized</u> and <u>only a few nonzero terms are contributed by each</u> <u>element</u>. In practice this localization results in <u>sparse equation systems</u>, reducing computer requirements. Whilst most desirable in practice this feature is not essential to the definition of the finite element process.

What then are the procedures by which a finite element approximation can be made? We have already mentioned—but now exclude from further discussion here—the *direct physical approach* and will concentrate on any problem which can be defined mathematically either by a (set of) differential equation(s) valid in a <u>domain Ω</u>

$$\mathscr{D}(\phi) = 0 \tag{1.3}$$

together with the associated boundary conditions on <u>boundaries (Γ)</u> of the domain

$$B(\phi) = 0 \tag{1.4}$$

---

* In some situations, such as for instance the behaviour of granular media, the first approach is still one of the most promising as continuously defined constitutive relations have not yet been adequately stated.

or by a variation principle requiring stationarity (max., min., or 'saddle') of some scalar functional Π

$$\Pi = \int_\Omega G(\phi)d\Omega + \int_\Gamma g(\phi)d\Gamma \qquad (1.5)$$

In both statements ϕ represents either the single unknown function or a set of unknown functions.

To clarify ideas, consider a particular problem presented by an electric field in a homogeneous dielectric medium where ϕ is the electric potential (a scalar quantity).

The specific governing equation is now written for a two-dimensional domain Ω as:

$$\mathscr{D}(\phi) \equiv \frac{\partial}{\partial x}\left(\epsilon \frac{\partial \phi}{\partial x}\right) + \frac{\partial}{\partial y}\left(\epsilon \frac{\partial \phi}{\partial y}\right) + \rho = 0 \qquad (1.6)$$

together with boundary conditions

$$B(\phi) = \phi_0 \text{ on } \Gamma_1$$

$$B(\phi) = \epsilon \frac{\partial \phi}{\partial n} \text{ on } \Gamma_2 \qquad (1.7)$$

in which both ϵ (the permittivity) and ρ (the volumetric charge density) may be functions of position and, in nonlinear problems, of the gradients or values of ϕ. (In the above, all the quantities are scalars.)

An alternative formulation (for linear problems) requires stationarity (a minimum) of a functional

$$\Pi = \int_\Omega \left\{ \frac{1}{2}\epsilon\left(\frac{\partial \phi}{\partial x}\right)^2 + \frac{1}{2}\epsilon\left(\frac{\partial \phi}{\partial y}\right)^2 - \rho\phi \right\} d\Omega - \oint_\Gamma \phi \frac{\partial \phi}{\partial n} \, d\Gamma \qquad (1.8)$$

for ϕ which satisfies only the first boundary condition.

In general if a functional Π exists then an associated set of (Euler) differential equations can always be found: the reverse is not necessarily true.

To obtain a finite element approximation to the general problem defined by Equations (1.3) or (1.5) we proceed as follows:

(a) the unknown function is <u>expanded</u> in a finite set of assumed, known <u>trial</u> <u>functions</u> N_i and unknown parameters a_i, i.e.,

$$\phi = \sum N_i a_i = \mathbf{Na}; \qquad (1.9)$$

(b) the approximation must be cast in a form of n equations which are defined as integrals over Ω and Γ, i.e.,

$$F_j = \int_\Omega E(\hat{\phi})d\Omega + \int_\Gamma e(\hat{\phi})d\Gamma \qquad j = 1 \text{ to } n. \qquad (1.10)$$

Immediately we note that the basic definitions of the finite element process previously given apply, as for integrable functions

$$\int_\Omega \left(\quad \right) d\Omega \equiv \sum \int_{\Omega^e} \left(\quad \right) d\Omega \tag{1.11}$$

and

$$\int_\Gamma \left(\quad \right) d\Gamma \equiv \sum \int_{\Gamma^e} \left(\quad \right) d\Gamma \tag{1.12}$$

in which Ω^e, Γ^e represent 'element' subdomains.

The problem of how the integrals of approximations are formed is thus the first, crucial, question of casting a problem in a finite element form.

1.3 APPROXIMATION INTEGRALS

1.3.1 Variational principles

If the problem is stated in terms of a stationary functional, then the formulation is most direct. We can write the approximate form of the functional as

$$\Pi \approx \hat{\Pi} = \Pi(\hat{\phi}) \tag{1.13}$$

and for stationarity we have a set of equations

$$F_j = \frac{\partial \Pi}{\partial a_j} = 0 \tag{1.14}$$

which by definition of Π is already cast in an integral form. This basis of forming a finite element approximation has been and remains most popular, providing a physically meaningful variational principle exists and can be readily identified. It is sometimes said that the finite element method is a 'variational process' which is, however, too limited a definition, as other alternatives, often more powerful, are present. The important question of how to proceed from the differential equation directly in cases where a variational principle does not exist or cannot be identified remains. The answer to it lies in the reformulation by use of weighting functions, or by the introduction of 'pseudo-variational' principles.

1.3.2 Weighted integral statements

It is obviously possible to replace the governing equations ((1.3, 1.4) or (1.6)) by an integral statement in all respects equivalent, i.e.,

$$\int_\Omega W^T \mathscr{D}(\phi) d\Omega + \int_\Gamma \bar{W}^T B(\phi) d\Gamma \tag{1.15}$$

in which W and $\bar{W}$ are completely arbitrary 'weighting' functions. Immediately, an approximation is possible in an integral form by choosing specific functions W_j and $\bar{W}_j$ and writing [3,4]

$$F_j = \int_\Omega W_j^T \mathscr{D}(\hat{\phi})\mathrm{d}\Omega + \int_\Gamma \bar{W}_j^T B(\hat{\phi})\mathrm{d}\Gamma. \qquad (1.16)$$

The process is known as the weighted residual method if $\mathscr{D}(\hat{\phi})$ and $B(\hat{\phi})$ are recognized as residuals by which the approximation misses the zero value required. Classical procedures of Galerkin's method, collocation, etc., are immediately recognized. The Galerkin process, in which the weighting function and the trial function are identical $(W_j \equiv N_j)$ is the most popular form used in finite element analysis.

Either form of deriving integral statements, and hence the set of approximating equations, can be and has been used in practice. The variational principle possesses, however, a unique advantage. If the function is quadratic in $\mathbf{a}$, the set of approximating Equations (1.14) can be written as

$$\mathbf{S}\mathbf{a} + \mathbf{P} = 0 \qquad (1.17)$$

in which $\mathbf{S}$ is always a symmetric matrix $(S_{ij}^T = S_{ji})$. For linear differential equations the weighting processes will also result in a similar set of equations via Equation (1.16): however, these will not in general be symmetric. The user of finite difference procedures may well be acquainted with such asymmetries, which often present computational difficulties. This symmetry can indeed be shown to be a precondition for the existence of a variational principle—it will be found that the Galerkin method of weighting will yield identical equations to those derived from a variation principle whenever this exists.

Because of this (and certain other) advantages of variational formulations, much work of a theoretical nature has been put in to establish equivalent functionals for problems defined by differential equations or to create pseudo-variation functionals.[5,6,7,8]

1.3.3 Pseudo-variational principles. Constraints by Lagrange multipliers or penalty functions. Adjoint variables and least square processes

Pseudo-variational principles can be established by various means. These include constrained variational principles and the extreme cases obtained from these by the use of adjoint functions, or the application of least square methods.

Constrained variational principles require that some functional Π be stationary and subject to constraints.

$$C(\phi) = 0. \qquad (1.18)$$

In such cases we can proceed to establish a new variational principle in either of two ways. In the first we introduce an additional set of functions λ known as *Lagrangian multipliers* and require

$$\bar{\Pi} = \bar{\Pi}\begin{Bmatrix} \phi \\ \lambda \end{Bmatrix} = \left(\Pi + \int_{\Omega} \lambda^T C d\Omega \right) \tag{1.19}$$

to be stationary. The variation of this functional results in

$$\delta\bar{\Pi} = \delta\Pi + \int_{\Omega} \delta\lambda^T C d\Omega + \int_{\Omega} \lambda^T \delta C d\Omega \tag{1.20}$$

which can only be true if both Π is stationary and the constraints (1.18) are satisfied.

The use of Lagrangian multipliers in practice is somewhat limited, owing to two drawbacks. First, the additional functions λ have to be discretized thus requiring a larger number of unknowns in the final problem. Second, it will always be found that, if Π is quadratic and C a linear equation, the final discretized form of Equation (1.17) has zero diagonal terms corresponding to the parameters discretizing λ (this is obvious from inspection of Equation (1.20)).

To obviate some of the difficulties associated with the use of Lagrangian multipliers it is possible to require the stationarity of a modified functional based on a penalty function. For, if at the solution we require a simultaneous satisfaction of the stationarity of Π and the satisfaction of constraints, we can minimize approximately

$$\bar{\Pi} = \Pi + \alpha \int_{\Omega} C^T C d\Omega \tag{1.21}$$

in which α is some large (positive) number 'penalizing' the error of not satisfying the constraints. As no procedure is without a drawback, we note here a purely numerical difficulty: as α becomes large the discretized equations tend to become ill-conditioned. However, with modern computers and high precision arithmetic, penalty function operations are becoming increasingly popular and their use more widespread.[9]

What if even a constrained variational principle does not appear to exist? Clearly both methods given above are still applicable by putting $\Pi \equiv 0$ and identifying the constraints with the full set of differential equations to be satisfied. Thus we can either make

$$\bar{\Pi} = \bar{\Pi}\begin{pmatrix} \phi \\ \lambda \end{pmatrix} = \int_{\Omega} \lambda^T C d\Omega \tag{1.22}$$

stationary, or alternatively minimize

$$\bar{\Pi} = \int_{\Omega} C^T C d\Omega. \tag{1.23}$$

The first is equivalent to the use of adjoint functions[10] while the latter is the straightforward application of the least squares procedures of approximation.[3,4]

The pseudo-variational principle established by Equation (1.22) in which a new, adjoint, function λ is introduced is of little practical use. The resulting discretized equation systems for parameters defining approximations to ϕ and λ are entirely uncoupled, and indeed there is little virtue in the symmetry arising from the whole system as a zero diagonal exists throughout. Nevertheless, this approach gives another interpretation of the Galerkin weighting process if similar expansions are used for ϕ and λ. The least square formulation, on the other hand, results in well-conditioned equation systems and deserves much wider attention in the finite element literature than it has so far received.[11,12]

1.3.4 Direct integral statements—virtual work

In many physical situations it is possible to formulate the problem directly in an integral form, avoiding the necessity of writing down the full governing differential equations. In particular the principle of virtual work in mechanics can be stated with greater generality than that arising from differential equations. Indeed, in such cases the weighted residual form given by Equation (1.15) arises in a form which can be obtained from such equations by the use of integration by parts. Such integration reduces the continuity requirement imposed on both functions W and N to 'integrability' (to which we shall refer in the next section). This relaxation of requirements is known mathematically as a 'weak formulation' of the problem. It is of philosophical interest to interject here a thought that perhaps such weak formulations are indeed the requirements of Nature as opposed to differential equations which, at certain physical discontinuities, are meaningless.

In structural mechanics, virtual work principles have almost replaced the formulation based directly on energy statements because of the wider applicability of virtual work (and because they often avoid complex algebraic manipulations).

Table 1.1 summarizes the basic processes by which the integral forms of approximation can be made as a preliminary to finite element analysis.

1.4   SEPARATION OF VARIABLES

At this point it is appropriate to mention that it is often convenient to discretize the problem only partially in a manner which, say, reduces a set of differential equations in three independent variables not directly to a numerical set of equations but to a lower order differential equation, say with only

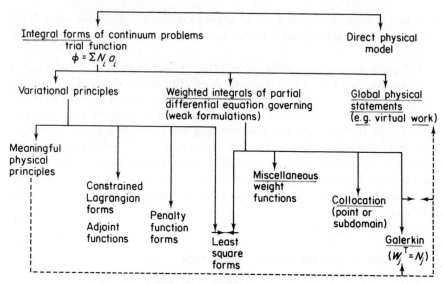

Table 1.1 Finite element approximation

one variable. This first differential equation can then, on occasion, be solved more efficiently by exact procedures or alternative numerical solutions.

Such a 'separation of variables' is particularly useful if the 'shape' of the domain in one of the independent directions is simple. This may arise if prismatic or axisymmetric shapes are considered in a three-dimensional problem or if one of the dimensions is that of time.

Considering the last case as a concrete example the trial function expression discretizing the unknown ϕ

$$\phi = \phi(x, \, y, \, z, \, t) \tag{1.24}$$

is made by modifying Equation (1.9) to

$$\phi = \sum N_i a_i = \mathbf{Na} \tag{1.25}$$

in which

$$N_i = N_i(x, \, y, \, z) \tag{1.26}$$

is only a function of position and a is a set of parameters which is a function of time

$$\mathbf{a} = \mathbf{a}(t). \tag{1.27}$$

'Partial variations' of variation principles (Equation (1.13)) or the use of any of the weighting procedures (Equation (1.16)) in which the weighting functions do not include the independent variable t can now be made, reducing the formulation to a set of ordinary differential equations.

We shall often find such a discretization useful and the ordinary set of differential equations can often be solved efficiently by simple finite difference schemes as well as by a secondary application of the finite element methodology.[1]

1.5 TRIAL FUNCTIONS

1.5.1 General principles

So far, beyond mentioning that the unknown function ϕ is expanded as in Equation (1.9) by a set of trial functions N, no specific mention was made of the form these trial functions should take or what limitations have to be imposed on them. We shall here consider, in very general terms, some of the guidelines. Details of the construction of trial functions for finite elements are described in Chapter 3.

As the trial functions N are constructed for practical reasons in a piecewise manner, using a different definition within each 'element', the question of required inter-element continuity is important in their choice. This continuity has to be such that either the integrals of the approximation given in general by Equation (1.10) or in particular forms by Equations (1.14) and (1.16) can be evaluated directly, without any contribution arising at the element 'interfaces'. Alternatively, such inter-element contributions must be of a kind that decrease continuously with the fineness of element subdivisions. The class of functions satisfying the first conditions shall be called conforming, whilst the ones which satisfy only the second one are named nonconforming (but usually admissible).

In general it is quite easy to specify the conformity conditions. If the integrand contains mth derivatives of the unknown functions ϕ, then the shape functions N have to be such that the function itself and its derivatives up to the order $m-1$ are required to be continuous (C^{m-1} continuity).[1]

In practice it is difficult to define conforming functions in a piecewise manner for any order of m greater than one and, because of this, many 'nonconforming' elements have originated in the past with the hope, sometimes proved *a posteriori*, that admissibility is achieved. The question of establishing admissibility is a difficult one and much work in this area is highly mathematical and not easy to interpret.[13,14,15] Simple tests of admissibility have, however, been devised and it is essential to subject any new nonconforming element to such an examination.[16,17]

There is, however, an intermediate position where inter-element contributions can be evaluated without the imposition of full conformity. This arises either if the derivatives of N occur in a linear form in the integrals and continuity can be relaxed by one further order,[18,19] or where in the basic formulation interface contributions are specifically inserted. The latter is the position with certain hybrid formulations [20,21] or Lagrange multiplier forms which specifically impose conformity as a constraint.[22,23] In this simple exposition we shall not be further concerned with these special situations, and will treat the conforming formulation as standard and the nonconforming one as a special variant of it.

A further condition which has to be imposed on shape functions is that of 'completeness', i.e. the requirement that in the limit, as the element size decreases indefinitely, the combination of trial functions should exactly reproduce the exact solution. This condition is simple to satisfy[1,17] if polynomial expressions are used in each element such that the complete mth order of polynomial is present, when mth order derivatives exist in the integral of approximations.

To demonstrate a few simple 'shape' functions we show in Figures 1.1 and 1.2 some piecewise defined shapes in one- and two-dimensional domains in which, to ensure C^0 conformity, the system parameter a_i takes on simply the value of the unknown function of certain points (often referred to as nodes) which are common to more than one 'element'. With this device a simple repeatable formula can be assigned to define N_i within any element.

It will be immediately recognized that as the parameter a_i influences the value of ϕ only in elements adjacent to a 'node' i, its contribution to the integrals will be limited to elements containing that node—hence the 'banded' feature of approximating equations already referred to.

It is of interest to note that such piecewise defined functions, which to many form the essence of the finite element method, were used for the first time in 1943 by Courant[24] despite the fact that integral approximation procedures in their general form are much older.

Today many complex forms of shape function definition exist, mostly developed in the last decade,[1] which are capable of being piecewise defined

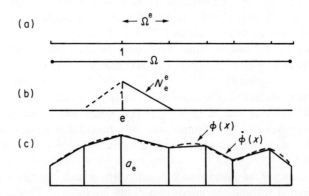

Figure 1.1   One-dimensional, simple, localized trial functions of C^0 continuity. (a) Domain subdivided into elements, (b) localized trial function for parameter $a_i = \phi_i$, (c) approximation to an arbitrary function $\phi(x)$

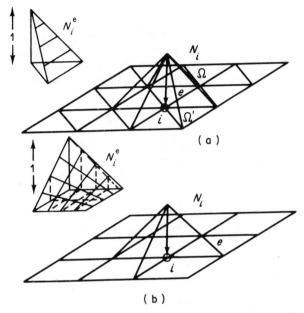

(a)

(b)

Figure 1.2 Two-dimensional localized trial functions of C^0 continuity. (a) Triangular element (linear expansion), (b) rectangular elements (bi-linear expansion)

and giving high orders of approximation. Some such functions are in fact defined, not in the simple original coordinate system in which the problem is given, but with a suitable transformation referred to curvilinear coordinates, by means of which most complex shapes of regions can be subdivided. Figure 1.3 shows some such elements of an 'isoparametric' kind[1,25] much used in practice. In all, C^0 continuity only is imposed and relatively simple formulation suffices.

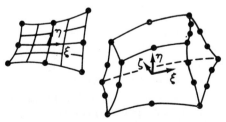

Figure 1.3 Some more elaborate elements with curvilinear coordinates

1.5.2 Particular example

To illustrate the process of discretization, which by this time must appear somewhat abstract to the reader, we shall return to the specific example given by Equation (1.6) and its associated boundary conditions, Equation (1.7).

Assuming that the potential ϕ (here a scalar quantity) can be approximated as

$$\hat{\phi} = \sum N_i a_i = \mathbf{Na} \tag{1.28}$$

in which both N_i and a_i are scalars and a_i is identified with nodal values of ϕ, we shall first use the variational principle of Equation (1.8).

Substituting Equation (1.28) into Equation (1.8) and differentiating with respect to a parameter a_j gives a set of equations required for stationarity as

$$F_j = \frac{\partial \Pi}{\partial a_j} = \frac{\partial}{\partial a_j} \left[\int \int_{\Omega} \left\{ \frac{\epsilon}{2} \left[\frac{\partial}{\partial x} \left(\sum N_i a_i \right)^2 \right] + \frac{\epsilon}{2} \left[\frac{\partial}{\partial y} \left(\sum N_i a_i \right)^2 \right] - \rho \sum N_i a_i \right\} dxdy \right.$$

$$= \int \int \left[\epsilon \frac{\partial N_j}{\partial x} \sum \frac{\partial N_i}{\partial x} a_i + \epsilon \frac{\partial N_j}{\partial y} \sum \frac{\partial N_i}{\partial y} a_i - \rho N_j \right] dxdy. \tag{1.29}$$

The whole equation system can be written as

$$\partial \hat{\Pi} = \mathbf{Sa} + \mathbf{P} = 0 \tag{1.30}$$

with

$$S_{ij} = \int \int_{\Omega} \epsilon \left[\frac{\partial N_i}{\partial x} \cdot \frac{\partial N_j}{\partial x} + \frac{\partial N_i}{\partial y} \cdot \frac{\partial N_j}{\partial y} \right] dxdy$$

$$P_j = - \int \int \rho N_j dxdy. \tag{1.31}$$

With the trial function assumed piecewise element by element it is simple to evaluate the integrals for each element, obtaining their contributions S_{ij}^e and P_j^e and obtaining the final equation by simple summation over all elements

$$S_{ij} = \sum S_{ij}^e$$
$$P_j = \sum P_j^e. \tag{1.33}$$

Alternative forms of approximation can be derived by the reader, using some weighting procedures described. It can be shown that in this linear case (i.e. in which ϵ and ρ are functions of position only) identical approximation will be available by application of the Galerkin weighting, but that other approximations will arise from use of alternative procedures. He will find that, for instance, application of least square processes to Equation (1.21) will result in second derivatives being present and will need C^1 continuity trial functions with subsequent difficulties of determining such functions. He will however

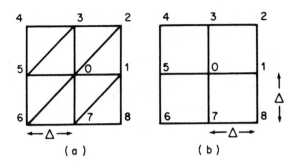

Figure 1.4 Regular triangle and square subdivisions

observe that the Galerkin process is available for nonlinear problems where the simple form of the variational principle is no longer applicable.

Using the approximation defined in Equations (1.30)–(1.31), it is of interest to insert a particular shape function and obtain in detail a typical discretized equation for a parameter j. Let us consider a typical internal node $j=0$ on a regular mesh of triangular elements as shown in Figure 1.4(a) in which a linear interpolation is used. Assuming that ϵ and ρ are constants and that the boundary does not occur in the vicinity, the contributions of all elements are found and coefficients S_{01}^e, S_{02}^e, etc. evaluated. After assembly, the typical equation becomes, after some algebraic manipulation,

$$\epsilon(\phi_1+\phi_3+\phi_5+\phi_7-4\phi_0)+\sigma\Delta^2=0. \tag{1.32}$$

The reader will recognize this as the standard finite difference equation obtained by direct point-differencing of Equation (1.6) and he may well inquire what advantage has been gained. Obviously, the numerical answers in this case are going to remain the same (at least if a boundary of the type Γ_2 does not occur). Immediately, however, it is important to point out that if ϵ varied discontinuously between elements (such as may be the case at interfaces between two regions of different permittivity) the finite element method would have yielded, in one operation, answers which direct finite difference procedures would tackle only by the introduction of additional constraints and interface conditions. Further, the variational form allows the gradient boundary condition to be incorporated directly.

Pursuing the problem further we try a rectangular element with a bilinear interpolation of ϕ as in Figure 1.4(b). The resulting finite element equation now becomes

$$\epsilon(\phi_1+\ldots+\phi_8)-8\epsilon\phi_0+6\rho\Delta^2=0 \tag{1.33}$$

a form substantially different from the standard finite difference equation which, although convergent to the same order of approximation, reduces the truncation error. Again the same comments can be made regarding the advantages of the finite element approximation.

In recent years much progress has been made in the finite difference methodology; in particular, integral forms including variational principles have been used as the basis of approximation in which the differences are only applicable to the differentials occurring in the integrals.[26-32] Comparison of such processes with finite element methods has been made by Pian[33] showing some detail of the problem discussed above. Such approaches eliminate some of the drawbacks of the finite difference procedures and indeed bring it close to the finite element process, as will be shown in the next section. However, the difficulty of increasing the order of approximation or of using irregular meshes still preserves the advantages of finite element processes.

1.6   THE RELATIONSHIP WITH OTHER NUMERICAL DISCRETIZATION PROCEDURES— FINITE DIFFERENCES—BOUNDARY SOLUTION METHODS

We have already remarked that several standard finite difference approximations can be derived using simple elements and Galerkin-type weighted formulations. The relative merits of the finite difference and finite element procedures are, therefore, not worthy of extensive discussion. Perhaps the only valid range of comparison concerns the questions of the use of simple, low order, versus more complex, high order, elements—the former being sometimes almost indistinguishable from classical finite differences. Here the situation is probelm dependent, and although higher order elements yield higher accuracy and convergence rates, in many applications the modelling needed for description of detail is such that the former prove advantageous.

While finite difference discretization is thus embraced by the general finite element form, the same is not so obviously true for another powerful numerical discretization technique—*the boundary solution method*—which has found considerable applicability and merit in electric and magnetic field problems. In this approximation use is made of the fact that for a range of linear and homogenous problems it is a relatively easy matter to generate a series of solutions which satisfy exactly the governing Equations (1.3) within the problem domain but which do not necessarily fit with the boundary conditions of Equation (1.4) required of the problem.

If the trial function is composed in the manner of Equation (1.9) of a series of such solution, the discrete parameters **a** have to be determined so that the boundary conditions are satisfied in an approximate manner. This requirement is generally met by a collocation procedure (which is but one of the weighting forms used in the generalized finite element process) and in this respect the boundary solution methods can again be considered as a *subclass of the finite* *element method in which the basic domain of discretization has been reduced to the boundary only.*

The advantage of boundary solution methods lies in

(*a*) the reduction of the problem domain (and hence, the number of unknown parameters) and

(*b*) the ease with which the trial functions can model singularities and infinite domains.

A special form of the boundary approximation formulates the problem in terms of an integral boundary equation to which the discretization procedure is applied. This technique is one of the most popular, and much literature has recently been devoted to the subject.[34]

As both the conventional finite elements and the boundary solution methods have their special merits, it appears desirable to combine their use in a single program in which various subdomains (elements) can be treated by the alternative approximations. In the context of application of the finite element methods to electromagnetic field problems this has been accomplished effectively by Silvester and Hsieh[35] and McDonald and Wexler[36] where the exterior, semi-infinite domain is substituted by a single element based on boundary integral methods. The process of linking and indeed formulating the boundary solution methods in a manner which is compatible with the standard finite element process has now been well developed[37,38] and the importance of such formulations which permit a single program to be used in a variety of modes is obvious.

Indeed, at this stage one can assert that the techniques are so well established and all-embracing that whenever a closed form solution is impossible the finite element method in its most general definition provides the best numerical answer.

1.7 AN EXAMPLE: 3-DIMENSIONAL MAGNETOSTATIC FIELDS

To conclude this chapter we invoke an example designed to illustrate not only the power of the finite element process, but also the importance of an intelligent formulation of the basic problem, the latter aspect being too often forgotten by over-enthusiastic followers.

The problem in question concerns the solution of three-dimensional magnetic fields in which the governing equations are

$$\nabla \times \mathbf{H} = \mathbf{J} \tag{1.34a}$$

$$\mathbf{B} = \mu \mathbf{H} \tag{1.34b}$$

$$\nabla \cdot \mathbf{B} = 0. \tag{1.34c}$$

Here **H**, **B**, and **J** are vector quantities (with three space components) denoting respectively the magnetic field strength, flux density, and current density. Only the latter is a prescribed quantity and the former present six problem unknowns. In the above μ stands for magnetic permeability and $\mathbf{V}$ for the operator

$$\mathbf{V} = \left[\frac{\partial}{\partial x} \hat{\mathbf{i}}, \frac{\partial}{\partial y} \hat{\mathbf{j}}, \frac{\partial}{\partial z} \hat{\mathbf{k}} \right] \tag{1.35}$$

and $\times$ for a vector product.

Assuming for the moment that μ is a quantity dependent on the position only, many possibilities of attack exist.

The most commonly used approach introduces a vector potential **A** defining

$$\mathbf{B} = \mathbf{V} \times \mathbf{A}. \tag{1.36}$$

This ensures the automatic satisfaction of Equation (1.34c) as

$$\mathbf{V} \cdot \mathbf{B} = \mathbf{V} \cdot (\mathbf{V} \times \mathbf{A}) = 0 \tag{1.37}$$

and on elimination of **H** one differential equation remains to be satisfied, namely

$$\mathbf{V} \times \frac{1}{\mu} \mathbf{V} \times \mathbf{A} = \mathbf{J}. \tag{1.38}$$

This equation is equivalent to a stationarity of a variational functional Π defined as

$$\Pi = \tfrac{1}{2} \int_{\Omega} \mathbf{A}^T \mathbf{V} \times \frac{1}{\mu} \mathbf{V} \times \mathbf{A} \ d\Omega - \int_{\Omega} A^T J d\Omega. \tag{1.39}$$

Using a Galerkin procedure on Equation (1.38) or performing the variation of (1.39) will lead to a discretization with the vector potential approximated in terms of *three* nodal parameters $\mathbf{a}_i$ as

$$\mathbf{A} = \sum N_i \mathbf{a}_i. \tag{1.40}$$

A difficulty arises in fully three-dimensional problems as, even with fully specified boundary values, **A** is not unique; Equation (1.37) being satisfied by a multiple of solutions

$$\mathbf{A}' = \mathbf{A} + \mathbf{V}\Psi \tag{1.41}$$

where Ψ are arbitrary functions. To overcome this an additional constraint is usually imposed, e.g. a requirement that

$$\mathbf{V} \cdot \mathbf{A} = 0. \tag{1.42}$$

While such a constraint can now be inserted into the variational form by procedures outlined in Section 1.3.3 the problem is unnecessarily complicated and an alternative is sought. Here we return to the basic equations and note that when μ is a constant throughout the domain an analytical solution can easily be written. For $\mu = 1$ such a solution takes the form of an integral

$$\Pi = \tfrac{1}{2} \int_{\Omega} \mathbf{A}^T \mathbf{V} \times \mathbf{V} \times \mathbf{A} d\Omega - \int_{\Omega} \mathbf{A}^T \mathbf{J} d\Omega \qquad (1.43)$$

where Ω is all space. If we now write

$$\mathbf{H} = \mathbf{H}_s + \mathbf{H}_m \qquad (1.44)$$

where $\mathbf{H}_m$ is any vector defined in such a way that

$$\mathbf{V} \times \mathbf{H}_m = \mathbf{J} \qquad (1.45)$$

then Equation (1.34a) becomes

$$\mathbf{V} \times \mathbf{H}_s = 0. \qquad (1.46)$$

This can be identically satisfied by the introduction of a scalar potential ϕ

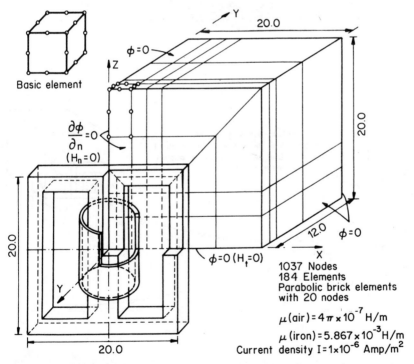

Figure 1.5 Three-dimensional magnetic field problem for transformer using scalar potential formulation showing the finite element mesh. 1037 nodes, 184 elements, parabolic brick elements with 20 nodes. $\mu(\text{air}) = 4 \times 10^{-7}$ H/m, $\mu(\text{iron}) = 5.867 \times 10^{-3}$ H/m. Current density $I = 1 \times 10^6$ Amp/m^2

defined by

$$\mathbf{H}_s = \mathbf{V}\phi. \tag{1.47}$$

Now the remaining set of equations (1.34b,c) yields, on eliminating **B**, a single governing differential equation

$$\mathbf{V} \cdot \mu \mathbf{V}\phi + \mathbf{V} \cdot \mu \mathbf{H}_m = 0 \tag{1.48}$$

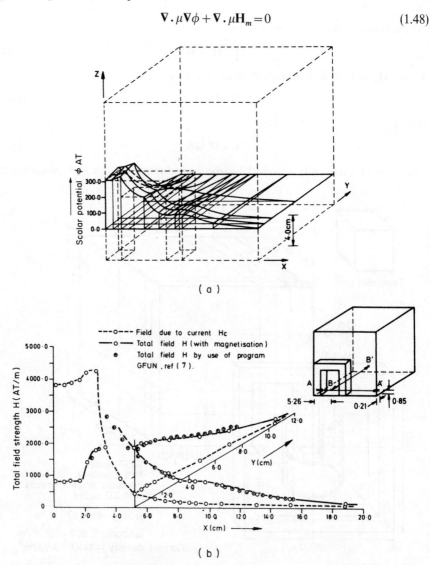

(a)

(b)

Figure 1.6 (a) Scalar potential on plane $z = 4.0$ cm. (b) Three-dimensional transformer example—field strength on lines AA′ and BB′. --○-- Field due to current $\mathbf{H}_s$.—○— Total field **H** (with magnetization)

with a *single scalar unknown* ϕ. Not only does the indeterminacy noted with the vector potential now disappear but, in a discretized form the number of unknowns is reduced to a third, a scalar expansion

$$\phi = \sum N_i a_i \tag{1.49}$$

being now used.

The detailed discretization of the above problem is identical to that used for Equation (1.8) if we note that

$$\mathbf{V} \cdot \mu \mathbf{\nabla} \phi = \frac{\partial}{\partial x}\left(\mu\frac{\partial \phi}{\partial x}\right) + \frac{\partial}{\partial y}\left(\mu\frac{\partial \phi}{\partial y}\right) + \frac{\partial}{\partial z}\left(\mu\frac{\partial \phi}{\partial z}\right) \tag{1.50}$$

and write the known quantity as

$$Q = \mathbf{V} \cdot \mu \mathbf{H}_m. \tag{1.51}$$

We see now that of the various possibilities the second provides a more economic approach, which indeed can be identified with a fairly standard problem for which many computer programs are available. This particular example is described by the author in detail elsewhere[39] and is illustrated here by a three-dimensional solution in Figures 1.5 and 1.6 in which three dimensional quadratic isoparametric elements are used.

For the reader of this book a moral of general applicability should be noted, i.e. that some thought on the nature of the analytical solution will invariably ease the subsequent numerical discretization. Further improvements of the scalar potential formulation have been reported recently by Trowbridge and Simpkin,[40] and these result in near optimal accuracy of solution as well as economy.

REFERENCES

1. Turner, M. J., Clough, R. W., Martin, H. C., and Topp, L. J. (1956). 'Stiffness and deflection analysis of complex structures', *J. Aero. Sci.*, **23**, 805–823.
2. Clough, R. W. (1960). 'The finite element in plane stress analysis', *Proc. 2nd ASCE Conf. on Electronic Computation, Pittsburgh, Pa., Sept. 1960.*
3. Crandall, S. H. (1956). *Engineering Analysis*, McGraw-Hill, New York.
4. Finlayson, B. A. (1972). *The Method of Weighted Residual and Variational Principles*, Academic Press, New York.
5. Oden, J. T. (1971). 'Finite element models of non linear operator equations', *Proc. 3rd Conf. on Matrix Meth. in Struct. Mech.*, Wright-Patterson AFB, Ohio.
6. Tonti, E. (1969). 'Variational formulation of non linear differential equations', *Bull. Acad. Roy. Belgique, Series 5*, **55**, 139–165, 262–278.
7. Sandhu, R. S. and Pister, K. S. (1971). 'Variational principles for boundary value and initial value problems in continuum mechanics', *Int. J. Solids Structures*, 7, 639–654.
8. Zienkiewicz, O. C. and Taylor, C. (1973). 'Weighted residual process in finite element method with particular reference to some transient and coupled problems',

In J. T. Oden and E. R. A. Oliveira (Eds), *Lectures on Finite Element Methods in Continuum Mechanics. Proc. NATO Symp. Lisbon*, Alabama Press.

9. Zienkiewicz, O. C. (1974). 'Constrained variational principles and penalty function methods in finite element analysis', *Conference on the Numerical Solution of differential equations, Univ. of Dundee, July 1973; Lecture Notes in Mathematics*, Springer Verlag, Berlin.

10. Arthurs, A. M. (1970). *Complementary Variational Principles*, Clarendon Press, Oxford.

11. Lynn, P. P. and Arya, S. K. (1975). 'Use of the least squares criterion in finite element formulation', *Int. J. Num. Meth. Eng.*, **6**, 75–88.

12. Zienkiewicz, O. C., Owen, D. R. J., and Lee, K. N. (1974). 'Least square finite element for elasto-static problems, use of reduced integration', *Int. J. Num. Meth. Eng.*, **8**, 341–358.

13. Zienkiewicz, O. C. (1977). *The Finite Element Method*, 3rd Edition, McGraw-Hill.

14. Oliveira, E. R. A. (1968). 'Theoretical foundations of the finite element method', *Int. J. Solids Structures*, **4**, 929–952.

15. Ciarlet, P. C. (1974). 'Conforming and non-conforming finite element methods for solving the plate problem', *Conference on the Numerical Solution of differential equations, University of Dundee, July 1973; Lecture Notes in Mathematics*, Springer Verlag, Berlin.

16. Bazeley, G. P., Cheung, Y. K., Irons, B. M., and Zienkiewicz, O. C. (1965). 'Triangular elements in bending-conforming and non-conforming solutions', *Proc. Conf. Matrix Method. Struct. Mech., Wright-Patterson AFB, Ohio*.

17. Strang, G. and Fix, G. J. (1973). *An Analysis of the Finite Element Method*, Prentice-Hall, Englewood Cliffs, N.J.

18. Nemat-Nasser, S. and Lee, K. N. (1973). 'Finite element formulations for elastic plates by general variational statements with discontinuous fields', *Report No. 39, The Technical University of Denmark*.

19. Nemat-Nasser, S. and Lee, K. N. (1973). 'Applications of general variational methods with discontinuous fields to bending, buckling and vibration of beams', *Comp. Methods in Appl. Mech. Eng.*, **2**, 1, 33–41.

20. Pian, T. H. H. and Tong, P. (1972). 'Finite element methods in continuum mechanics', *Advances in Applied Mechanics*, Vol. 12, Academic Press, New York.

21. Tong, P. (1970). 'New displacement hybrid finite element model for solid continua', *Int. J. Num. Methods Eng.*, **2**, 78–83.

22. Harvey, J. W. and Kelsey, S. (1971). 'Triangular plate bending element with enforced compatibility', *AIAA J.*, **9**, 6, 1023–1026.

23. Szabo, B. A. and Tsai, C. T. (1973). 'The quadratic programming approach to the finite element method', *Int. J. Num. Meth. Eng.*, **5**, 375–381.

24. Courant, R. (1943). 'Variational methods for the solution of problems of equilibrium and vibration', *Bull. Am. Math. Soc.*, **49**, 1–23.

25. Zienkiewicz, O. C., Irons, B. M., Ergatoudis, J., Ahmad, S., and Scott, F. C. (1969). 'Isoparametric and associated element families for two- and three-dimensional analysis'. In I. Holand and K. Bell (Eds), *Finite element method in stress analysis*, Tapir, Trondheim, Norway.

26. Courant, R. and Hilbert, P. (1953). *Methods of Math. Physics*, Vol. 1, Interscience, New York.

27. Forsythe, G. E. and Wasow, W. R. (1960). *Finite Difference Methods for Partial Differential Equations*, Section 20, Wiley, New York.

28. Young, D. M., Jr. (1972). 'Survey of numerical analysis'. In J. Todd (Ed), *The numerical solution of elliptic and parabolic partial differential equations*, Addison-Wesley, New York.

components of the field vectors must be continuous, except if a surface current exists (as may be the case for a superconductive medium). In the latter case, the magnetic field vector may possess a discontinuity equal to the surface current density $\mathbf{J}_s$,

$$\mathbf{n} \times (\mathbf{E}_1 - \mathbf{E}_2) = 0 \qquad (2.5)$$

$$\mathbf{n} \times (\mathbf{H}_1 - \mathbf{H}_2) = \mathbf{J}_s. \qquad (2.6)$$

Similarly, it follows from the Maxwell divergence equations that the normal components of both flux densities must be continuous at any material interface, except if a surface charge should reside at the interface. In that case, the electric flux density vector may possess a discontinuity equal to the surface charge density. Thus

$$\mathbf{n} \cdot (\mathbf{D}_1 - \mathbf{D}_2) = \sigma \qquad (2.7)$$

$$\mathbf{n} \cdot (\mathbf{B}_1 - \mathbf{B}_2) = 0. \qquad (2.8)$$

No other boundary or interface conditions can be deduced from the Maxwell equations. As a result, these equations together with the interface continuity conditions describe an infinitely-extending field, hardly convenient from a computational viewpoint. It is usual to bound an electric or magnetic field problem by introducing surface current densities and surface charge densities σ such as to force the appropriate field or flux density vectors to vanish on a convenient boundary. For example, let it be assumed that material 1 in Figure 2.1 is perfectly conductive. It can then be argued from energy conservation principles that no volume charge distribution and no volume current density can exist in that material, hence that $\mathbf{H}_1$ and $\mathbf{D}_1$ must vanish. Equations (2.6) and (2.7) thereby establish a relationship between $\mathbf{H}_2$ and $\mathbf{J}_s$ on the one hand, and $\mathbf{D}_2$ and σ on the other. These boundary conditions are sufficient to allow solution of the bounded problem.

The equations as sketched out so far are applicable regardless of the nature of the materials involved. However, they are computationally quite intractable, not only because there are four partial differential equations to be solved simultaneously, but also because the system of equations still remains indeterminate; there are four equations, but five distinct vectors. Since two of the differential equations are vector equations, the implied total number of (scalar) differential equations is in fact eight, the number of independent vector components fifteen. Additional equations are obtained by relating the flux densities to the fields through the constitutive equations, which in fact amount to descriptions of the material media,

$$\mathbf{D} = \epsilon \mathbf{E} \qquad (2.9)$$

$$\mathbf{B} = \mu \mathbf{H}. \qquad (2.10)$$

Electromagnetic problems, in general, may be divided into various classes, depending on the nature of the permittivity and the permeability. In general, these quantities may be nonlinear tensor quantities. Fortunately, in a substantial number of practical cases they are nearly constants, and for many other materials they may be nonlinear but scalar. A similar classification of materials is possible with respect to the conductivity g, which relates the current density to the electric field through Ohm's law:

$$\mathbf{J} = g\mathbf{E}. \tag{2.11}$$

The number of equations thus assembled is sufficient to determine the field vectors. However, from a computational point of view the direct solution of Maxwell's equations is rarely to be recommended. Direct substitution of Equations (2.9)–(2.11) still leaves the two field vectors $\mathbf{E}$ and $\mathbf{H}$ to be determined, thus at every space point six independent components must be computed. Solutions in this form have occasionally been carried out, but it is usual to seek out more economic representations.

2.3 THE WAVE EQUATIONS

One common method of reducing computational complexity in practical problems is to replace the above first-order differential equations, involving at least two independent vector variables, by differential equations of second order but involving only one vector. This procedure is advantageous not only because it reduces the number of variables, but also because operators of even order are very convenient to handle by the finite element method. Such equations are usually called wave equations because certain of their solutions represent travelling or standing waves. To demonstrate how they may be arried at, consider the Maxwell magnetic curl equation (2.2). Taking curls on both sides, one obtains

$$\text{curl curl } \mathbf{H} = \text{curl } \mathbf{J} + \frac{\partial}{\partial t} \text{ curl } \mathbf{D}. \tag{2.12}$$

Let attention be concentrated for the moment on linear, isotropic materials, i.e. materials which possess constant scalar permittivity and conductivity. Substitution of Equations (2.9) and (2.11) yields

$$\text{curl curl } \mathbf{H} = g \text{ curl } \mathbf{E} + \frac{\partial}{\partial t} \epsilon \text{ curl } \mathbf{E}. \tag{2.13}$$

However, curl $\mathbf{E}$ may be replaced by an equivalent magnetic vector quantity by substituting Equation (2.1). There results

$$\text{curl curl } \mathbf{H} = -g\frac{\partial \mathbf{B}}{\partial t} - \epsilon\frac{\partial^2 \mathbf{B}}{\partial t^2}. \tag{2.14}$$

Finally, substitution of (2.10) produces a single equation stated entirely in terms of the magnetic field vector **H**:

$$\text{curl curl } \mathbf{H} = -\mu g \frac{\partial \mathbf{H}}{\partial t} - \mu \epsilon \frac{\partial^2 \mathbf{H}}{\partial t^2}. \tag{2.15}$$

While this form is entirely satisfactory, it is conventional to rewrite it, using the identity

$$\text{curl curl } \mathbf{H} = \text{grad div } \mathbf{H} - \nabla^2 \mathbf{H} \tag{2.16}$$

which is valid for any differentiable vector. Making use of this identity, Equation (2.15) may be rewritten as

$$\nabla^2 \mathbf{H} - \mu g \frac{\partial \mathbf{H}}{\partial t} - \mu \epsilon \frac{\partial^2 \mathbf{H}}{\partial t^2} = \text{grad div } \mathbf{H}. \tag{2.17}$$

However, in accordance with Equation (2.4), the divergence of the magnetic flux density vanishes identically everywhere. Thus, the right-hand side of Equation (2.17) must vanish, leaving finally

$$\nabla^2 \mathbf{H} - \mu g \frac{\partial \mathbf{H}}{\partial t} - \mu \epsilon \frac{\partial^2 \mathbf{H}}{\partial t^2} = \mathbf{0}. \tag{2.18}$$

This equation, valid in any linear isotropic material, is generally termed the *wave equation* in **H**.

A similar result may be derived for the electric field **E**, by first taking curls of both sides of Equation (2.1), then proceeding to substitute in much the same manner as above. The development is generally very similar, with the exception that the divergence of the electric flux density vector **D** does not necessarily vanish. As a result, the wave equation obtained in **E** is inhomogeneous:

$$\nabla^2 \mathbf{E} - \mu g \frac{\partial \mathbf{E}}{\partial t} - \mu \epsilon \frac{\partial^2 \mathbf{E}}{\partial t^2} = \frac{1}{\epsilon} \text{grad } \rho. \tag{2.19}$$

This equation, like its homogeneous counterpart in **H** above, lends itself to finite element techniques rather better than the Maxwell equations themselves. Nevertheless, it is still far from ideal for most purposes, since it involves four independent variables (three space dimensions and time) so that four-dimensional finite elements are required. These are quite unavoidable in some classes of problems; however, for many engineering applications, further simplifications are possible. Some of these will be discussed below.

2.4 PROBLEMS WITH TIME-INVARIANT BOUNDARIES

There exist many physical problems in which relative motion does not occur, i.e. in which all boundaries and interfaces in the problem remain stationary in time. In such cases, separation of space and time coordinates becomes possible. The single second-order differential Equation (2.18) or (2.19), with four independent variables, is then replaced by a system of two equations, one dealing with space variations only, the other with time only.

The process involved is classical. The supposition is initially made that the field vector involved, for example the magnetic field **H** of Equation (2.18), is representable as a product of a function $T(t)$ of time only, and a vector function $S(x, y, z)$ of the space coordinates only,

$$\mathbf{H}(x, y, z, t) = \mathbf{S}(x, y, z)\, T(t). \tag{2.20}$$

Substitution of this form into (2.18), and division by T, then produces

$$\frac{T\mathbf{V}^2\mathbf{S}}{T} - \frac{\mu g \mathbf{S} T'}{T} - \frac{\mu \epsilon \mathbf{S} T''}{T} = \mathbf{0}. \tag{2.21}$$

The division by T is proper, for T is a scalar function, continuous, and does not vanish (except possibly on a set of measure zero). Equation (2.21) may be rewritten, grouping terms, as

$$\mathbf{V}^2\mathbf{S} - \left(\frac{\mu g T' + \mu \epsilon T''}{T}\right)\mathbf{S} = \mathbf{0}. \tag{2.22}$$

The classical variable separation argument may now be invoked. On inspection, it is clear that only the factor in parentheses in Equation (2.22) has anything to do with time, for **S** is a function of the space coordinates only. Thus, if (2.22) is to hold at all values of time, the quantity in parentheses must necessarily be a constant, say $-k^2$. That is to say, there exists some constant $-k^2$ such that

$$\mu \epsilon \frac{\mathrm{d}^2 T}{\mathrm{d}t^2} + \mu g \frac{\mathrm{d}T}{\mathrm{d}t} + k^2 T = 0 \tag{2.23}$$

and that

$$(\mathbf{V}^2 + k^2)\mathbf{S} = \mathbf{0}. \tag{2.24}$$

The latter is generally known as the vector Helmholtz equation, for it involves the Helmholtz differential operator and a vector variable **S**. On the other hand, Equation (2.23) is a straightforward second-order ordinary differential equation with constant coefficients, for which analytic solutions are well-known.

The variable separation argument can only lead to successful problem solutions if the boundary conditions as well as the differential Equation (2.18) can be separated. In other words, the position of each boundary or interface involved in the problem must be independent of time; motion of boundaries is not permitted. There are clearly large numbers of problems for which this supposition is true, and for which the variable separation process will therefore yield a useful reduction in problem complexity. Of course, in situations where all significant problem boundaries are parallel to one or more coordinate axes (problems with translational or rotational symmetry, for example), further variable separation may still be possible, simplifying the problem even further.

2.5   MAGNETIC AND ELECTRICAL POTENTIALS

A fundamental result of vector analysis is that the divergence of the curl of any twice differentiable vector **A** vanishes identically,

$$\text{div curl } \mathbf{A} = 0. \tag{2.25}$$

The Maxwell magnetic divergence equation (2.4) is thus satisfied if one defines the vector **A**, commonly known as the *magnetic vector potential*, such that

$$\mathbf{B} = \text{curl } \mathbf{A}. \tag{2.26}$$

Of course, **A** is not fully defined by (2.26), since the Helmholtz theorem of vector analysis states that a vector is uniquely defined if and only if both its curl and divergence are known, as well as its value at some one space point. The latter added datum is necessary because a uniform vector, with identically the same value everywhere in space, has both zero curl and divergence, so that specifying curl and divergence of a vector specifies the vector only to within an additive uniform vector.

If a magnetic vector potential **A** is defined in any manner which satisfies Equation (2.26), the Maxwell electric curl Equation (2.1) may be written in the form

$$\text{curl}\left(\mathbf{E} + \frac{\partial \mathbf{A}}{\partial t}\right) = \mathbf{0}. \tag{2.27}$$

It would be tempting to conclude from this equation that the electric field **E** is the negative of the time rate of change of **A**, but such a conclusion would not be true. Analogously to (2.25), the curl of any gradient vanishes, so that for any twice differentiable scalar function V,

$$\text{curl grad } V = \mathbf{0}. \tag{2.28}$$

Thus the only conclusion to be drawn from Equation (2.27) is that **E** differs from the time rate of **A** by some irrotational vector grad V,

$$\mathbf{E} = -\frac{\partial \mathbf{A}}{\partial t} - \text{grad } V. \tag{2.29}$$

Usually, V is known as the *electric scalar potential*, because in the static case when no time variations are present, the electric field becomes merely the negative gradient of V:

$$\mathbf{E}_{\text{static}} = -\text{grad } V. \tag{2.30}$$

It is immediately evident that electrostatic field problems are often best studied by employing the electric scalar potential V. Being a scalar quantity, V allows the fewest possible independent data to be associated with every finite element node. Indeed, nearly all electric field analysis reported in the literature to date are carried out in terms of the electric scalar potential.

The magnetic and electric potentials are also useful in time-varying situations, in which case they obey certain (possibly modified) wave equations. A wave equation in $\mathbf{A}$ may be derived by taking curls of both sides of Equation (2.26), obtaining

$$\text{curl curl } \mathbf{A} = \text{curl } \mathbf{B}. \tag{2.31}$$

Substitution of the magnetic constitutive Equation (2.10), and subsequently of the Maxwell magnetic curl Equation (2.2), yields in the case of homogeneous and isotropic materials

$$\text{curl curl } \mathbf{A} = \mu\left(\mathbf{J} + \frac{\partial \mathbf{D}}{\partial t}\right). \tag{2.32}$$

The electric flux density vector $\mathbf{D}$ may be eliminated from this equation by substituting the electric constitutive relationship (2.9) and Equation (2.29). There results

$$\nabla^2 \mathbf{A} = \text{grad}\left(\text{div } \mathbf{A} + \mu\epsilon\frac{\partial V}{\partial t}\right) + \mu\epsilon\frac{\partial^2 \mathbf{A}}{\partial t^2} - \mu\mathbf{J}. \tag{2.33}$$

Since the divergence of $\mathbf{A}$ has not so far been specified anywhere, this equation can be simplified very considerably by choosing the divergence of $\mathbf{A}$ so as to satisfy the so-called *Lorentz condition*

$$\text{div } \mathbf{A} = -\mu\epsilon\frac{\partial V}{\partial t}. \tag{2.34}$$

Thus there finally emerges the inhomogeneous wave equation

$$\nabla^2 \mathbf{A} - \mu\epsilon\frac{\partial^2 \mathbf{A}}{\partial t^2} = -\mu\mathbf{J}. \tag{2.35}$$

In the analysis of magnetostatic fields, the wave equation in $\mathbf{A}$ assumes simply the form of a vector Poisson equation,

$$\nabla^2 \mathbf{A} = -\mu\mathbf{J}. \tag{2.36}$$

For such fields, this wave equation does not provide any particular advantage over the wave Equation (2.18) in $\mathbf{H}$, since both are vector equations. However, there do occur many circumstances, particularly where translational or rotational symmetry exists, in which $\mathbf{J}$ possesses only one vector component. In such cases, $\mathbf{A}$ may possess only one component also, and (2.35) or (2.36) become quasi-scalar equations. Thus, in two-dimensional magnetostatics the vector potential **A** frequently provides the same computational advantages as does the electric scalar potential V in electrostatic problems.

It should be emphasized that the Lorentz choice is not the only possible one for the divergence of **A**, nor even always desirable. Returning to Equation (2.32), the two rightmost terms may be eliminated by substituting the constitutive relations (2.9) and (2.11), then inserting Equation (2.29). These substitutions yield the rather complicated-looking result

$$\mathbf{V}^2\mathbf{A} = \text{grad div } \mathbf{A} + \text{grad}\left(\mu g V + \mu\epsilon\frac{\partial V}{\partial t}\right) + \mu g \frac{\partial \mathbf{A}}{\partial t} + \mu\epsilon\frac{\partial^2 \mathbf{A}}{\partial t^2} \qquad (2.37)$$

which is easily reduced into a very simple form, however, by choosing the divergence of **A** appropriately:

$$\text{div } \mathbf{A} = -\mu g V - \mu\epsilon\frac{\partial V}{\partial t}. \qquad (2.38)$$

With this choice, **A** obeys the homogeneous wave equation

$$\mathbf{V}^2\mathbf{A} - \mu g \frac{\partial \mathbf{A}}{\partial t} - \mu\epsilon\frac{\partial^2 \mathbf{A}}{\partial t^2} = \mathbf{0}. \qquad (2.39)$$

This choice occasionally proves convenient in low-frequency problems, in which the second time derivative term may be neglected.

Finally, another common choice for the divergence of **A** is simply

$$\text{div } \mathbf{A} = 0 \qquad (2.40)$$

usually known as the *Coulomb convention*. As may be seen from Equations (2.33) and (2.37), this choice yields quite complicated equations except in the static case where all time derivatives vanish. Its use is therefore generally confined to static field problems.

2.6 TWO-DIMENSIONAL STATIC AND DYNAMIC FIELDS

As an example of the manner in which magnetic vector potential problems may occur in translationally symmetric geometries, consider an electric machine rotor conductor placed in a slot, as indicated in Figure 2.2. It will be assumed that the conductor is nonmagnetic, while the surrounding magnetic material is infinitely permeable. The current carried by the conductor flows entirely in the longitudinal (say z) direction so that the magnetic vector potential possesses only a z-directed component:

$$\mathbf{A} = \mathbf{1}_z A. \qquad (2.41)$$

As a consequence, the vector Poisson Equation (2.36) now assumes an

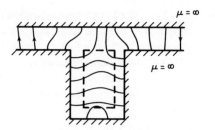

Figure 2.2 Electric machine slot-
conductor

apparent scalar form:

$$\nabla^2 A = -\mu J. \tag{2.42}$$

Although the current density J in the conductor is known, this equation can of course not be solved without a suitable set of boundary conditions. In the present context, with the iron surfaces assumed infinitely permeable, magnetic flux lines must impinge on each iron surface at right angles. In the air gap itself, at a sufficient distance from the slot, it can reasonably be assumed that a flux line crosses the gap in a straight-line fashion. It is not difficult to show that in two-dimensional cases such as the present one, every flux line corresponds to a contour of constant A, and conversely. Hence, the boundary conditions applicable to Figure 2.2 and Equation (2.42) are

$$\frac{\partial A}{\partial n} = 0 \qquad \text{at iron surfaces}$$

$$\frac{\partial A}{\partial n} = 0 \qquad \text{at the symmetry line} \tag{2.43}$$

$$A = 0 \qquad \text{along some flux line.}$$

If a time-varying current flows in the conductor, the above formulation is, of course, no longer valid, for the fields are not static any more. The boundary conditions sketched out above remain applicable, but Equation (2.39) rather than (2.36) applies. However, the problem can still be reduced to a scalar form in exactly the same manner as above, Equation (2.41); currents and vector potentials everywhere still remain unidirectional. Since time variations in a problem of this nature are likely to be fairly slow, displacement current in the conductor is usually negligible as compared to conduction current,

$$\mu\epsilon\frac{\partial^2 \mathbf{A}}{\partial t^2} \ll \mu g\frac{\partial \mathbf{A}}{\partial t}. \tag{2.44}$$

After reduction to two dimensions and elimination of the negligible higher derivative term, Equation (2.39) in this particular case thus assumes the form

$$\mathbf{V}^2\mathbf{A} - \mu g \frac{\partial \mathbf{A}}{\partial t} = \mathbf{0}. \qquad (2.45)$$

This equation is homogeneous, and does not contain the conductor current. An alternative formulation, based on the Lorentz convention, could be obtained so as to exhibit the current density $\mathbf{J}$ explicitly. However, such a formulation would not be of much use since in the time-varying case the current density distribution depends on the magnetic field, which is only known once $\mathbf{A}$ has been calculated.

To make the formulation of Equation (2.45) computationally useful, let $\mathbf{A}$ be written as the sum of two portions $\mathbf{A}_0$ and $\mathbf{A}_e$,

$$\mathbf{A} = \mathbf{A}_0 + \mathbf{A}_e \qquad (2.46)$$

where $\mathbf{A}_0$ represents the solution of the corresponding static problem, Equations (2.42) and (2.43). That is, $\mathbf{A}_0$ represents the magnetic vector potential distribution that would exist were the current variations in time to be very slow. Since the static problem as formulated above is readily solvable, $\mathbf{A}_0$ may be considered known. Substitution of (2.46) into Equation (2.45), however, produces

$$\mathbf{V}^2\mathbf{A}_e - \mu g \frac{\partial \mathbf{A}_e}{\partial t} = -\mathbf{V}^2\mathbf{A}_0 + \mu g \frac{\partial \mathbf{A}_0}{\partial t} \qquad (2.47)$$

subject to the same boundary conditions (2.43) as are applicable to the static problem. It may be noted that the first term on the right of Equation (2.47) is in fact the static current density distribution, multiplied by permeability, since $\mathbf{A}_0$ is the solution of the static problem (2.42).

In the case of sinusoidal time variation, the problem may be recast in complex form, so that time derivatives no longer appear. There results in that case

$$\mathbf{V}^2\mathbf{A}_e - j\omega\mu g \mathbf{A}_e = \mu \mathbf{J}_0 + j\omega\mu g \mathbf{A}_0 \qquad (2.48)$$

where $\mathbf{J}_0$ represents the current distribution that would exist in the absence of time variation, i.e. in the low-frequency limit.

2.7   RETARDED POTENTIALS

Many electromagnetic field problems, particularly in antenna theory but also in many other applications areas, involve fields in infinitely extending free space. The analysis of such situations is frequently best handled through the so-called *retarded potentials*.

As shown above, the vector potential $\mathbf{A}$ satisfies a wave equation; it can easily be shown that the electric scalar potential V does likewise. Suppose it is desired to find the scalar potential V at any arbitrary point P in the neighbourhood of a point charge q placed at the point Q in infinitely extending free space. Since free space is lossless, the potential V satisfies the lossless wave equation

$$\mathbf{V}^2 V - \mu\epsilon \frac{\partial^2 V}{\partial t^2} = 0 \tag{2.49}$$

everywhere except at the point Q itself. This very simple problem can be solved analytically. Recasting (2.49) in spherical coordinates centred on Q, there is obtained

$$\frac{1}{r^2} \frac{\partial}{\partial r}\left(r^2 \frac{\partial V}{\partial r}\right) - \mu\epsilon \frac{\partial^2 V}{\partial t^2} = 0. \tag{2.50}$$

This equation is readily solved by making the variable substitution $U = rV$, which results in

$$\frac{\partial^2 U}{\partial r^2} - \mu\epsilon \frac{\partial^2 U}{\partial t^2} = 0 \tag{2.51}$$

which is satisfied by any solution of the form

$$U = h(t \pm r\sqrt{(\mu\epsilon)}) \tag{2.52}$$

where $h(z)$ is some twice differentiable function. Thus, the electric scalar potential will be given as a wave of some waveshape, travelling radially toward or away from the charge q:

$$V = \frac{h(t \pm r\sqrt{(\mu\epsilon)})}{r}. \tag{2.53}$$

For small r, this solution clearly should reduce to the electrostatic potential of a point charge q. This electrostatic potential of a slowly varying charge $q(t)$ is given by

$$\lim_{r \to 0} V = \frac{h(t)}{r} = \frac{q(t)}{4\pi r\epsilon} \tag{2.54}$$

permitting identification of the so far arbitrary function $h(z)$, and thus yielding the solution

$$V = \frac{1}{4\pi\epsilon r} q(t - r\sqrt{(\mu\epsilon)}). \tag{2.55}$$

It will be noted that the positive sign has been discarded. It is usually argued that the potential V exists as a consequence of the existence of the charge $q(t)$, so that an outward travelling potential wave originating from events at the point Q is reasonable. On the other hand, it is thought unreasonable to believe

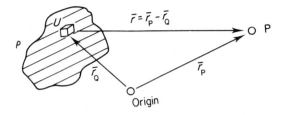

Figure 2.3 A general spatial charge distribution may be thought of as a collection of volume charge elements ρdU

that a radially inward-travelling wave of potential causes the charge to exist. V is commonly known as a retarded potential, for it associates the instantaneous potential at a distant point P to the value of the charge $q(t)$ not at the same instant, but at some previous time: the time delay or retardation being equal to the length of time required by the wave to travel the distance from Q to P. It is evident from Equation (2.55) that the speed of travel is $1/\sqrt{(\mu\epsilon)}$, commonly denoted by c and equal to the velocity of light.

The result derived above for the point charge can immediately be widened to apply to much more general cases. Suppose the point Q is located in a finite-sized charge cloud (e.g. in a current-carrying conductor), so that the charge residing at Q is $\rho(t)dU$, where ρ is the local volume charge density and dU represents the volume element, as in Figure 2.3. Clearly the potential V at point P is influenced by events at Q to the extent of a potential contribution

$$dV = \frac{1}{4\pi\epsilon r}\,\rho(t-r/c)dU. \tag{2.56}$$

Summing over all of the space occupied by charges, the potential at P is given by

$$V = \int \frac{1}{4\pi\epsilon r}[\rho]dU. \tag{2.57}$$

The square brackets in this equation are conventionally employed to indicate time retardation. That is, $[\rho]$ is a representation for $\rho(Q, t-r/c)$, i.e. the value of volume charge density at point Q at a time sufficiently retarded to allow for the propagation time from Q to P. In a similar way, the magnetic vector potential **A** associated with a current distribution **J** can be shown to be a retarded potential,

$$\mathbf{A} = \int \frac{\mu}{4\pi r}[\mathbf{J}]dU. \tag{2.58}$$

Equations (2.57) and (2.58) are physically simple and comprehensible, but mathematically very complicated to use, since time retardation is not easy to handle in most cases. However, the special case of sinusoidally time-varying quantities is relatively simple. Suppose, for example, that the value of volume charge density at Q is given by the real part of

$$\rho = \rho_0(x,\ y,\ z)e^{j\omega t}. \tag{2.59}$$

The retarded value of volume charge density is then

$$[\rho] = \rho_0(x,\ y,\ z)e^{j\omega(t - r/c)} \tag{2.60}$$

which may immediately be rewritten in the form

$$[\rho] = \rho e^{-jkr} \qquad k = \omega/c \tag{2.61}$$

showing that in this particular form of time-variation, time retardation is equivalent to a phase shift. Thus, the general, but mathematically unwieldy, retarded potential Equation (2.57) may be rewritten in the special case of sinusoidal time-variation as

$$V = \int \frac{1}{4\pi\epsilon} \frac{e^{-jkr}}{r} \rho dU \tag{2.62}$$

where no time retardation is implied in the integrand. Correspondingly, the magnetic vector potential equation becomes

$$\mathbf{A} = \int \frac{\mu}{4\pi} \frac{e^{-jkr}}{r} \mathbf{J} dU. \tag{2.63}$$

These equations are Fredholm integral equations of the first kind. They represent complete descriptions of physical problems; no additional boundary conditions are required to make them solvable. In this respect, they are frequently simpler to treat than differential equations.

2.8   INTEGRAL EQUATIONS FOR LOW FREQUENCY FIELDS

Formulation of the equations of electromagnetics in integral form is traditional in antenna theory, but also convenient in any problem area where essentially unbounded spaces are encountered. Consider, for example, the static electric field surrounding a charged object in free space. The electric potential V in its neighbourhood is correctly given by Equation (2.62). However, the wavenumber k assumes zero value for static fields, so that the retarded potential Equation (2.62) assumes the very simple form

$$V = \int \frac{1}{4\pi\epsilon r} \rho dU \tag{2.64}$$

where the volume of integration encompasses all space containing charges. If the charged object under consideration is a conductor, only surface charges may exist. In this case, the integration covers only the surface of the object,

$$V = \oint \frac{1}{4\pi\epsilon r} \, \boldsymbol{\sigma} \cdot d\mathbf{S}. \tag{2.65}$$

This equation, and others related to it, have been used in recent years to solve a variety of static field problems. The integral equation formulation is particularly convenient here, for although the potential V is defined in an infinite space region, a finite element discretization is necessary only for those portions of space on which the integral operator acts; i.e. all that is required is a finite element model of the conductor surface.

For a long cylindrical conductor of uniform cross-section, the surface element $d\mathbf{S}$ may be written in terms of the longitudinal (z-directed) and transverse distance measures,

$$d\mathbf{S} = \mathbf{1}_z dz \times \mathbf{1}_t dn. \tag{2.66}$$

Integration with respect to z may be carried out immediately; (2.65) then assumes the form,

$$V = \frac{1}{2\pi\epsilon} \oint (\sigma \Delta z) \log r \, dn \tag{2.67}$$

the remaining integration being carried out over the conductor perimeter in the transverse plane. It will be noted that the integrand in this case has a logarithmic singularity, while in the three-dimensional cases above a singularity of the form r^{-1} occurs. The only significant difficulty in finite element treatment of integral operators is the numerical integration around such singularities. Methods for treating both the above types of singularity have been extensively investigated, and appear in the literature.

Magnetic potential fields yield to similar treatment. Taking the current density components in Cartesian coordinates one at a time, each leads to a result similar to (2.64); combining them, there is obtained

$$\mathbf{A} = \frac{\mu}{4\pi} \int \frac{1}{r} \, \mathbf{J} dU \tag{2.68}$$

for the magnetic vector potential $\mathbf{A}$. Where translational symmetry obtains, as in the case of a long current-carrying conductor of uniform cross-section, only the longitudinally directed current component of $\mathbf{J}$ and, correspondingly, only the longitudinally directed component of $\mathbf{A}$ exist. Direct integration in the longitudinal direction then yields a result analogous to Equation (2.67)

$$\mathbf{A}_z = \frac{\mu}{2\pi} \int \mathbf{J}_z \, \log r \, dS. \tag{2.69}$$

The latter case is particularly interesting in that the magnetic field **H** and the vector potential **A** are both physically vector quantities; however, **A** possesses only one component while **H** has two. Thus, a considerable complexity reduction in computation is achieved through the use of **A**.

In analysing magnetic fields, the choice of using the field **H** or the vector potential **A** is frequently resolved in favour of **A**, even if the boundary conditions applicable to **A** are more complicated in form. In the general case where both vectors have three space components, no complexity reduction occurs of the type encountered in (2.69) above. Nevertheless, **A** remains more attractive when finite element models are employed. Given any approximation to **A** constructed by finite element methods, the corresponding magnetic field **H** is uniquely defined and physically realizable, since it is derived from **A** by the curl operator and is therefore always nondivergent. On the other hand, if **H** is modelled by a finite element approximation, guaranteeing nondivergence is much more difficult. In the latter case, it is usual to find a multiplicity of solutions. From these the correct one will have to be selected by seeking the most nearly nondivergent member of that linear manifold of all fields **H** which approximately satisfy the Maxwell equations. This process is often difficult.

2.9   SUMMARY AND CONCLUSIONS

In the present state of the finite element art, numerous electric and magnetic field problems have been solved, and numerous others remain to be solved. Two-dimensional problems resulting from translational or rotational symmetry are both fairly easy to formulate and are computationally tractable, so that most of the work to date has concentrated on solving these. As computers grow in capacity, and the price of computation decreases, it is to be expected that more and more three-dimensional problems will be attempted. Efforts to do so have been made in a number of areas, but usually only with qualified success. A major difficulty lies in the construction of suitable finite element models without data blunders. Another major stumbling block is the interpretation of results. In two-dimensional field solutions, graphical displays of results are considered almost essential, and in many respects they are even more so in three dimensions. Unfortunately, the representation of three-component vector fields in a three-dimensional space in an easily interpretable and comprehensible fashion is still an unsolved problem. Hence the cost of three-dimensional solutions will, for the foreseeable future, remain high in terms of human labour, even though the computational cost is decreasing rapidly. Consequently, the formulation of field equations in two-dimensional or quasi-two-dimensional special cases will, for the foreseeable future, continue to occupy a major role in analysis.

Finite Elements in Electrical and Magnetic Field Problems
Edited by M. V. K. Chari and P. P. Silvester
© 1980, John Wiley & Sons Ltd.

Chapter 3

Shape Functions

R. H. Gallagher

3.1 INTRODUCTION

The finite element concept, as defined in Chapter 1, involves the division of the total domain of a problem into subdomains—the finite elements—and the description of the behaviour of each of the subdomains by functions expressed in terms of a limited number of parameters. Although such functions might be chosen of a relatively complicated form, it has been customary to employ simple polynomial expansions. The choice of functions is clearly a major aspect of the finite element method and the purpose of this chapter is to review established approaches to their selection.

In the period of early development of the finite element method, in the 1950's, the selection of behaviour functions for the various types of elements under development was a haphazard affair. Functions were often chosen to correspond to an intuitive feeling for the response of the class of problem to be modelled. The conditions which all functions ought to meet were not identified, nor were organized schemes to establish functions for any desired geometric form and number of node points. Melosh,[1] in 1961, and later Bazeley, et al.[2] established such conditions and schemes. The latter have been more sharply defined and expanded in the intervening years.

Rather detailed summaries of functions which describe the behaviour of finite elements are given by Prenter,[3] Mitchell and Wait,[4] and the present writer,[5] among others. In this chapter we outline the more fundamental aspects of certain of these representations. The representation of elements can be classified in many ways: by geometric form, by the character of the functions employed (polynomial, trigonometric, etc.), and by the conditions that the functions are required to meet.

The usual geometric forms of planar elements are triangular and quadrilateral. Csendes, in Chapter 7, gives a fairly complete account of triangular elements, so in this chapter we concentrate on quadrilaterals. The simpler case of the one-dimensional element and the more general case of the three-dimensional element are included. Attention is concentrated upon polynomial functions. The conditions to be met by such functions principally concern the order of continuity required at physical and inter-element boundaries. The usual finite element formulations of electric and magnetic field problems demand the interelement continuity of the approximating functions themselves, and not of their derivatives. Consequently, we emphasize polynomial functions which meet the former but not the latter conditions.

We begin this chapter with a discussion of the properties of polynomial functions which are relevant to finite element approximation. Then the notion of a shape function is introduced and implemented for linear, planar rectangles, and three-dimensional elements. The important concept of isoparametric mapping, which enables the construction of elements with curved or distorted sides, is described. This leads naturally, in the closing segments, to a method of characterizing singular behaviour.

3.2   POLYNOMIAL FUNCTIONS

Polynomial functions give a powerful means for the description of complex behaviour and, simultaneously, lend themselves well to the processes of integration and differentiation. In one dimension the general polynomial expansion can be written as

$$\phi = a_1 + xa_2 + x^2 a_3 + \ldots + x^m a_n \tag{3.1}$$

where $a_0, \ldots, a_n$ denote the parameters of the expansion and x is the coordinate dimension. In one dimension the total number of parameters, n, and the order of the polynomial, m, are related by $n = m + 1$. We can express this in matrix form as

$$\phi = \boldsymbol{a}\boldsymbol{x} = \boldsymbol{x}\boldsymbol{a} \tag{3.2}$$

where

$$\boldsymbol{a} = [a_1 \, a_2 \, a_3 \ldots a_n] \tag{3.3}$$

$$\boldsymbol{x} = [1 \, \text{x} \, \text{x}^2 \ldots x^m] \tag{3.4}$$

and $\boldsymbol{a} = \boldsymbol{a}^T$, $\boldsymbol{x} = \boldsymbol{x}^T$, where the superscript T denotes the transpose of the matrix to which it is assigned.

We have, in two dimensions, for complete mth order polynomial with n terms

$$\phi = a_1 + xa_2 + ya_3 + x^2 a_4 + xya_5 + y^2 a_6 + \ldots + a_n y^m \tag{3.5}$$

and, by analogy with the development in one dimension, we write

$$\phi = \boldsymbol{x}, \boldsymbol{y} \, \boldsymbol{a} = \boldsymbol{a} \, \text{x}, \boldsymbol{y} \tag{3.6}$$

or, using the summation convention

$$\phi = \sum_{i=1}^{n} x^j y^k a_i. \tag{3.7}$$

The superscripts j and k are integer exponents whose values are related to the integer subscript i as follows

$$i = \tfrac{1}{2}(j+k)(j+k+1) + k + 1. \tag{3.8}$$

All combinations of j and k must be taken, in ascending order, from 0 to m.

As stated above, the order (m) of the polynomial is the highest power to which a variable is raised in the chosen expansion and the expansion is *complete* to a given order if it contains all terms of that order. If a polynomial is complete to order m we will designate the vector of variables as $x, y(m)$. Thus, for such a case we write Equation (3.6) as

$$\phi = x, \, y(m) \, a = a \, x, \, y(m). \tag{3.6a}$$

The number of terms, n, in a complete polynomial in two dimensions, is

$$n = \tfrac{1}{2}(m+1)(m+2). \tag{3.9}$$

As an example, for a first order expansion

$$\phi = a_1 + a_2 x + a_3 y = x, \, y(1) \, a. \tag{3.10}$$

Here, Equation (3.9) confirms that $n = 3$.

It is convenient, in dealing with various aspects of two-dimensional polynomial series, to refer to the Pascal triangle format in arraying the terms of the expansion. Thus

a_1	(constant—1 term)
$a_2 x \quad a_3 y$	(linear—2 terms)
$a_4 x^2 \quad a_5 xy \quad a_6 y^2$	(quadratic—3 terms)
$a_7 x^3 \quad a_8 x^2 y \quad a_9 xy^2 \quad a_{10} y^3$	(cubic—4 terms)
$a_{11} x^4 \quad a_{12} x^3 y \quad a_{13} x^2 y^2 \quad a_{14} xy^3 \quad a_{15} y^4$	(quartic—5 terms).

This for simplicity is given up to the fifth order polynomial, but the extension to any order is apparent. The array shows, at a glance, the number of terms present in a complete polynomial of a given order and the number of terms associated with an individual order, e.g. there are 21 terms in a complete quintic order polynomial and 6 terms of fifth order.

The term at which a polynomial series approximation of element behaviour should be truncated deserves close attention. It would at first appear that an element geometric form is selected and its number of vertices defines the needed number of terms in a polynomial expansion. The next section demonstrates, however, that it is possible to define elements of any desired level of complexity with node points at almost any desired location. Thus, the problem can be viewed as one of first choosing a level of approximation, defined by the number of terms in the polynomial expansion, and then proceeding to selection of an element/nodal point arrangement.

The above remarks imply that the number of terms of a polynomial series and the number of node points on the element are identical. It is, in fact, possible to choose either a fewer or a greater number of terms in the series than there are node points. Such circumstances represent special cases and will not be discussed here.

The polynomial series must, in general, include the constant term (a_1) and the linear terms $(a_2 x, a_3 y)$. The constant term and a combination of the linear terms enable the definition of a fixed frame of reference for the problem. Without these terms and, therefore, without the ability to specify a fixed frame or reference, the system of algebraic equations resulting from the finite element idealization will be singular.

The functional (or integral relationship upon which the finite element representation is based) will be comprised of first derivatives in those cases where second order differential equations govern the problem. After insertion of the polynomial into the functional, and after differentiation, constant terms must be present. This must be the case because the value of the functional must approach a constant as the element size is reduced. It follows that all linear terms must be present since these yield the constants after differentiation.

The most noteworthy condition to be met in the choice of a polynomial for approximation of behaviour within the element is that which has become known as the inter-element continuity condition. This condition arises from the requirement for a unique evaluation of the functional identified with the problem under analysis. Continuity of all derivatives up to one order less than the maximum order appearing in the functional is needed. The satisfaction of this condition has presented some of the more formidable difficulties in the construction of behaviour functions in finite element analysis. Procedures are available for accomplishing this objective for simple elements, however, and these are developed later in this chapter.

The concept which will be exploited subsequently in the establishment of inter-element–continuous polynomial representations can be outlined here in rather simple terms. If the description of a function along an inter-element boundary is specified uniquely by points along that boundary, then inter-element continuity of that function is preserved in the joining of the element to the adjacent element. For example, if a cubic function is employed to describe the displacement of an element, four independent points must be available on an edge for the purpose of describing the function on that edge. With this in mind we can establish one criterion for the number of terms to be chosen in the polynomial series.

According to Equation (3.9), the total number of terms available in a complete mth-degree two-dimensional polynomial is $\frac{1}{2}(m+1)(m+2)$. Consider an R-sided plane polygonal element wherein ϕ is to be defined by the polynomial series in terms of its values at specified locations on only the boundary of the element. For unique definition on each boundary segment, it must be described by $(m+1)$ points, and a total of $R(m+1) - R = Rm$ such points will be required on the boundary of the complete polygon. Equating the avilable number of coefficients to the required number, we have

$$\frac{1}{2}(m+1)(m+2) = Rm \qquad (3.11)$$

a condition that is met by only $R = 3$ and $m = 1$ or $m = 2$. These elements are the three-noded and six-noded triangles. The present chapter is devoted, for the most part, to consideration of four-sided elements. It is clear that for such elements it is not possible to employ polynomial series that are complete to a given order with node points that lie solely on the periphery of the element. Various artifices have been employed to surmount this limitation, such as reducing the order of integration of the functional in which the polynomial will be used,[6] but in the routine development of element algebraic relationships this limitation prevails.

It is important to observe that polynomial series can be defined for an element in a form that violates inter-element compatibility conditions. Formulations of this type have been widespread in finite element analysis, especially in the early years of development of the method. The resulting formulations may prove to be entirely satisfactory in practical application, although in certain cases it may prove otherwise. To confirm the validity of such 'nonconforming' formulations one can apply the 'patch test'.[7-9] This is a simple numerical experiment in which a patch of element is subjected to an imposed displacement state which should produce a state within the elements wherein the derivatives of the variable ϕ are constant throughout the patch. Failure to realize this state indicates the possibility that the formulation will not yield, in the limit, correct solutions to the physical problem being represented. We limit our attention, in this chapter, to 'conforming' representations.

3.3 SHAPE FUNCTIONS

3.3.1 The Shape-function concept

The discussion of the properties of polynomial functions given in the preceding section dealt entirely with polynomials written in terms of 'generalized parameters', a_i. In dealing with the physical problem, however, it is necessary to express the approximating function in terms of the nodal point values of the physical variables, e.g. potential or flux. In this section we examine the methods of definition of polynomials in such terms and study the character of the resulting expressions.

We refer to the nodal point values of the physical variables as 'degrees-of-freedom'. Such measures of behaviour can be the variable itself or its derivatives of various orders. There can be significant computational advantages in the use of derivatives as nodal point degrees-of-freedom, but these may be overshadowed by their awkwardness in application. These and other pros and cons of derivatives as nodal point degrees of freedom are discussed in Reference 5. Here, we limit attention to the use of the variable itself as the nodal point degree-of-freedom.

Thus, we now consider the establishment and properties of expression of the following form

$$\phi = N_1\phi_1 + N_2\phi_2 + \ldots + N_i\phi_i + \ldots + N_n\phi_n$$

$$= \sum_{i=1}^{n} N_i\phi_i = \mathbf{N}\boldsymbol{\phi} \tag{3.12}$$

where ϕ is the set of node point parameters, ϕ_i is the ith node point parameter (degree-of-freedom) of the element, N_i is the 'shape function' corresponding to ϕ_i, and n is the total number of degrees-of-freedom at the node points of the element. A shape function, which is given physical meaning below, is a function of the spatial coordinates of the element.

Consider, for simplicity, one-dimensional behaviour (Figure 3.1). Here, Equation (3.12) becomes

$$\phi = N_1\phi_1 + N_2\phi_2. \tag{3.13}$$

A linear polynomial expression is consistent with the availability of two degrees-of-freedom, so we adopt the function

$$\phi = a_1 + a_2 x. \tag{3.14}$$

The constants a_1 and a_2 have, as yet, no identified physical meaning. Their meaning, and the desired form of the expression for the independent variable, are obtained by evaluation of Equation (3.14) at the boundary points and solution of the resulting equations. This gives $a_1 = \phi_1$ and $a_2 = (\phi_2 - \phi_1)/L$, and Equation (3.14) becomes

$$\phi = \left(1 - \frac{x}{L}\right)\phi_1 + \frac{x}{L}\phi_2. \tag{3.15}$$

Hence $N_1 = (1 - x/L)$ and $N_2 = x/L$.

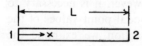

Figure 3.1   One-dimensional element

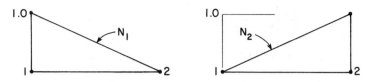

Figure 3.2   Shape functions for one-dimensional element

Figure 3.2 shows the variation of N_1 and N_2 as a function of the element coordinate x. Each plot describes the shape of the element when the element is given a unit value at the node point corresponding to N_i (point 1 in the case of N_1, point 2 in the case of N_2) and the value of ϕ at the other degree of freedom is held to zero. It describes the displaced shape of ϕ within the element under these circumstances, hence the designation *shape function*.

It is instructive to examine N_1 and N_2 more closely. Firstly, they are nondimensional, are unity when evaluated at the nodal point to which they refer, and are zero when evaluated at the other nodal point. Thus, $N_1 = 1$ at $x = 0$, which gives $\phi = \phi_1$, as it must be. Such properties hold for the shape functions of all types of elements in which the function is described by its own values at the node points. Their definition must be extended somewhat when derivatives of the function are employed as degrees of freedom.

The nondimensional property also deserves attention. If the element geometry is characterized by the coordinate ξ which varies from 0 to 1 in the interval between the end points, then $\xi = x/L$ and $N_1 = (1 - \xi)$, $N_2 = \xi$. The origin of ξ could, alternatively, be placed at the centre of the interval 0–1. We would then have $\xi = 2x/L - 1$ and $N_1 = \frac{1}{2}(\xi - 1)$, $N_2 = \frac{1}{2}(\xi + 1)$. These are different forms of 'natural' coordinates. Our preference, in the work which follows, will be for physical coordinates which are positioned at an extreme corner of the element. Subsequently, in Section 3.4, we will use natural coordinates which have their origin at the centroid of the 'unit length' $(-1, 0, +1)$, 'unit square', or 'unit cube'.

Now we proceed to construct shape functions directly. We can avoid the lugubrious task of writing and solving equations defined initially in terms of the parameters $a_1, \ldots, a_n$ through the use of Lagrangian interpolation concepts, which give the coefficients (in this case, the shape functions) of a polynomial series passing through a specified number of points. The formula for Lagrangian interpolation is[5]

$$N_i = \frac{\displaystyle\prod_{\substack{j=1 \\ j \neq i}}^{m+1} (x - x_j)}{\displaystyle\prod_{\substack{j=1 \\ j \neq i}}^{m+1} (x_i - x_j)} \tag{3.16}$$

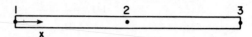

Figure **3.3** Three-noded one-dimensional
element

where the symbol Π denotes a product of the indicated binomials $((x-x_j)$ or $(x_i-x_j))$ over the indicated range of j.

In expanded form, this is

$$N_i = \frac{(x-x_1)(x-x_2) \dots (x-x_{m+1})}{(x_i-x_1)(x_i-x_2) \dots (x_i-x_{m+1})}. \tag{3.17}$$

For three points in a line, Figure 3.3, we have $m=2$, $x_1=0$, and $x_3=2x_2$. Thus

$$\phi = \frac{(x-x_2)(x-2x_2)}{2x_2^2}\phi_1 + \frac{x(2x_2-x)}{x_2^2}\phi_2 + \frac{x(x-x_2)}{2x_2^2}\phi_3. \tag{3.18}$$

3.3.2   Planar rectangular elements

One-dimensional finite element formulations are of limited value. Indeed, one of the principal motivations for the development of the finite element method was to produce a practical tool for the analysis of two- and three-dimensional continua. The Lagrangian interpolation approach generalizes easily to the construction of shape functions for two- and three-dimensional behaviour. We illustrate this generalization by means of rectangular elements; as noted previously, a treatment of triangular domains is given by Csendes in Chapter 7.

The simplest rectangular element is one based upon node points at the vertices only (Figure 3.4). The selected number of node points is associated with a bilinear field, written as follows

$$\phi = N_{1x}N_{1y}\phi_1 + N_{2x}N_{1y}\phi_2 + N_{2x}N_{2y}\phi_3 + N_{1x}N_{2y}\phi_4 \tag{3.19}$$

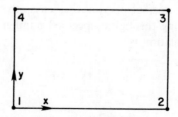

Figure 3.4   Four-noded
rectangle

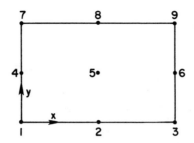

Figure 3.5 Nine-noded rectangle

where $N_{1x}=(1-\xi)$, $N_{2x}=\xi$, $N_{1y}=(1-\eta)$, $N_{2y}=\eta$. ξ and η are the x- *and* y-direction nondimensional coordinates, with $\xi=x/x_2$ and $\eta=y/y_3$. Note that each shape function is the product of independent x-only and y-only shape functions. Furthermore, in defining the x-direction behaviour the side connecting points 1 and 4 is treated as the left ('point 1') end of a one-dimensional element and the side connecting points 2 and 3 is treated as the right end (see Figure 3.1). The y-direction interpolants are similarly defined, based on sides 1–2 and 3–4, respectively.

One can proceed to the rectangle with mid-side node points and an interior node, Figure 3.5, and establish the shape functions consistent with a 'biquadratic' field:

$$\phi=(N_{1x}N_{1y})\phi_1+(N_{2x}N_{1y})\phi_2+(N_{3x}N_{1y})\phi_3+(N_{1x}N_{2y})\phi_4+(N_{2x}N_{2y})\phi_5$$
$$+(N_{3x}N_{2y})\phi_6+(N_{1x}N_{3y})\phi_7+(N_{2x}N_{3y})\phi_8+(N_{3x}N_{3y})\phi_9 \qquad (3.20)$$

where $N_{1x}=[(x-x_2)(x-x_2)]/2x_2^2$ is the node 1 shape function for x-direction quadratic Lagrangian interpolation, derived previously (see Equation 3.18), and N_{1y}, N_{2x}, N_{2y}, etc., are similarly defined.

The full interpolation of a quadratic function results in an interior node point. Such interior points are regarded as a nuisance and it is customary to eliminate them from the expression for ϕ before making use of the expression in finite element formulations. Procedures for eliminating interior nodes are detailed in Reference 5. *(primary condensation ?)*

The character of the functions obtained in the above procedure and their correspondence to particular terms of a polynomial expansion can be established by reference to the Pascal triangle, Figure 3.6. First, note that a complete one-dimensional polynomial of a given order corresponds exactly to a Lagrange interpolant of the same order. Thus, the first order polynomial $\phi=a_1+a_{2x}$ corresponds to the interpolant $\phi=(1-x/L)\phi_1+(x/L)\phi_2$. Bilinear interpolation, in terms of generalized coordinates, is given on the Pascal triangle by the product of first order functions in x and y. This encompasses four terms, so we have

$$\phi=a_1+a_2x+a_3y+a_5xy. \qquad (3.21)$$

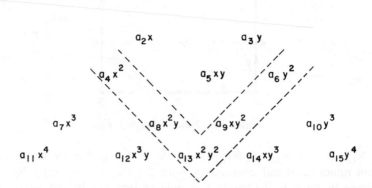

Figure 3.6   Pascal triangle—an array of the polynomial series in
two dimensions

This corresponds to the four-term Lagrange interpolant given by Equation (3.19).

The polynomial terms which are encompassed by a higher order 'bivariate' (two-dimensional) Lagrangian interpolant are now easily identified by means of this scheme. The quadratic bivariate expansion is the product of the quadratic expansions in x and y, respectively, and gives 9 terms. These nine terms, in the Lagrange interpolant form, have been written as Equation (3.20).

The foregoing illustrates that bivariate Lagrange interpolation on the rectangle involves, for a polynomial of degree n, a total of $(n+1)^2$ terms. These encompass all terms of an nth order polynomial in two dimensions; this is designated as a polynomial 'complete' to the order n. The additional terms extend to a term of order $2n$. For example, the first order interpolant is complete to the first degree (a_1, $a_2 x$, $a_3 y$) and extends to the second degree term $a_5 xy$. Since convergence rates are governed by the highest order of complete polynomial a full expansion in two dimensions on the rectangle might not produce a solution of the best accuracy for the given number of nodes employed.

3.3.3   Solid elements (hexahedra)

Generalization of the above ideas to three-dimensional elements of hexahedral form (Figure 3.7) presents no difficulty and will be commented upon briefly.

The basic 8-noded hexahedron, with nodes only at the vertices, corresponds to Lagrangian interpolation of linear functions in the three directions. We obtain

$$\phi = (N_{1x}N_{1y}N_{1z})\phi_1 + (N_{2x}N_{1y}N_{1z})\phi_2 + \ldots + (N_{1x}N_{1y}N_{2z})\phi_8 \qquad (3.22)$$

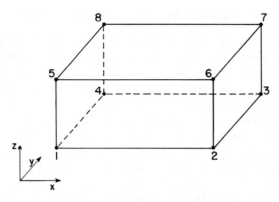

Figure 3.7 Eight-noded hexahedron

where now ξ is the nondimensional coordinate in the z-direction and the shape function pairs (N_{1x}, N_{2x}), (N_{1y}, N_{2y}) and (N_{1z}, N_{2z}) are those which are obtained through one-dimensional interpolants in the x, y, and z directions, respectively.

In progressing to the quadratic representation the element nodal point array shown in Figure 3.8a, consisting of 27 node points, is realized in consequence of trivariate Lagrangian interpolation. This representation, however, is extremely awkward on account of the nodal point at the centroid of the element as well as those at the middle of the faces. Consequently, the standard practice is to employ a 20-point element, of the configuration shown in Figure 3.8b, which has node points only at the vertices and along the edges. The relevant shape functions, which can be derived by operations on the complete (27-node) Lagrangian interpolation and which preserve inter-element continuity of ϕ, are given in Reference 5.

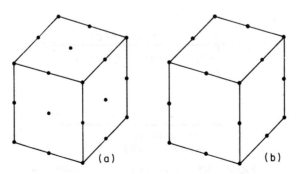

Figure 3.8 Hexahedra for quadratic fields

3.4 ISOPARAMETRIC FORMULATIONS

The term 'isoparametric element' refers to the construction of formulation wherein the geometry of the element is described by means of the same ('iso') parameters as are used in the description of the behaviour of the element. It is necessary that the 'conforming' shape functions be employed as these parameters, or else gaps will be present along the inter-element boundaries in the geometric description of the region to be analysed. The important advantage of isoparametric representation is the ease with which it can be used to describe curved boundaries by means of quadratic, cubic, or even higher-order polynomial-based shape functions. For this reason isoparametric representation has become by far the most popular method of geometric description, although it is by no means the only approach available. (For some alternatives see References 10–12.)

We outline, in the following, the basic notions of isoparametric representation. Some special properties of isoparametric representation which are useful in the presence of singular behaviour are reserved for discussion in the next section.

Isoparametric representation depends upon a particular manner of description of the element shape functions. To clarify the issues involved we first consider the shape functions of a three-noded one-dimensional element. This element was introduced in Figure 3.3 and is repeated here in Figure 3.9. It is described in physical coordinates by an x-coordinate axis aligned with the axis of the element, with origin positioned at point 1 on the left end of the member. The points are numbered 1, 2, and 3; the intermediate point, 2, is in general not midway between the ends of the element. The element is described in terms of a nondimensional, natural coordinate ξ in Figure 3.9a. This coordinate has its origin midway between the end points $\xi = -1$ and $\xi = +1$. Thus, the relationship between x and ξ is

$$\xi = \frac{2x - L}{L}.$$ (3.23)

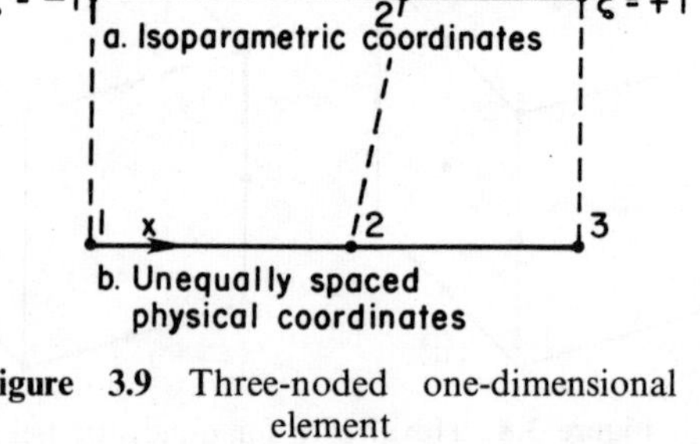

[handwritten margin notes: † (strictly)]

[handwritten margin notes: † but if not So, then bias can easily become a singularity — See later]

a. Isoparametric coordinates

b. Unequally spaced physical coordinates

Figure 3.9 Three-noded one-dimensional element

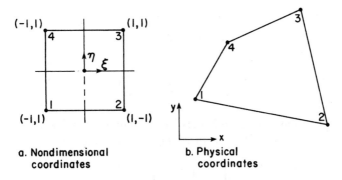

Figure 3.10 Four-noded planar isoparametric element

Now, if we substitute this expression into the previously established Lagrangian interpolation formula, we have

$$\phi = -\frac{\xi}{2}(1-\xi)\phi_1 + (1-\xi^2)\phi_2 + \frac{\xi}{2}(1+\xi)\phi_3 \qquad (3.18a)$$

and it is clear that the shape functions refer only to nondimensional coordinates and have no relationship to the physical location of the node points.

With the above consideration in mind we progress to the treatment of the simplest two-dimensional element, the four-noded element shown in Figure 3.10b. Figure 3.10a illustrates the unit square in the nondimensional coordinates ξ and η. We have for shape functions, upon adaptation of Equation (3.19) (note that ξ is defined here as $(2x - L)/L$, whereas it was x/L in Equation (3.19)),

$$N = \frac{1}{4} \lfloor (1-\xi)(1-\eta) \quad (1+\xi)(1-\eta) \quad (1+\xi)(1+\eta) \quad (1-\xi)(1+\eta) \rfloor. \quad (3.24)$$

We use these shape functions to describe not only ϕ but also the x and y coordinates of points on the element,

$$x = N\mathbf{x}, \, y = N\mathbf{y} \qquad (3.25)$$

where $\mathbf{x}$ and $\mathbf{y}$ list the x and y coordinates of the node points of the element. In this case

$$\mathbf{x} = \lfloor x_1 \; x_2 \; x_3 \; x_4 \rfloor^T \qquad (3.26)$$

and similarly for $\{\mathbf{y}\}$.

To construct an element matrix we must find the first derivatives of ϕ with respect to x and y. The ϕ, however, are functions of the ξ and η coordinates. Hence, we must find a relationship between the derivatives with respect to x and y and the derivatives with respect to ξ and η. By the chain rule of

* Author may be thinking of $F = -\nabla \phi$.

differentiation,

$$\frac{\partial N_i}{\partial \xi} = \frac{\partial x}{\partial \xi}\frac{\partial N_i}{\partial x} + \frac{\partial y}{\partial \xi}\frac{\partial N_i}{\partial y}$$

$$\frac{\partial N_i}{\partial \eta} = \frac{\partial x}{\partial \eta}\frac{\partial N_i}{\partial x} + \frac{\partial y}{\partial \eta}\frac{\partial N_i}{\partial y}.$$ (3.27)

(handwritten: $\left\lfloor \frac{\partial}{\partial \xi_j} \right\rfloor = \left\lfloor \frac{\partial}{\partial x_i} \right\rfloor \left[\frac{\partial x_i}{\partial \xi_j} \right]$ $\left\{ \frac{\partial}{\partial \xi_j} \right\} = \left[\frac{\partial x_i}{\partial \xi_j} \right]^{\mathsf{T}} \left\{ \frac{\partial}{\partial x_i} \right\}$)

We can determine $\partial x/\partial \xi = (\partial N/\partial \xi)x$ by differentiation of Equation (3.25) and similarly for $\partial x/\partial \eta$, etc. We can therefore write Equation (3.27) as

$$\begin{Bmatrix} \dfrac{\partial N_i}{\partial \xi} \\ \dfrac{\partial N_i}{\partial \eta} \end{Bmatrix} = [\mathscr{J}]^{\mathsf{T}} \begin{Bmatrix} \dfrac{\partial N_i}{\partial x} \\ \dfrac{\partial N_i}{\partial y} \end{Bmatrix}$$ (3.28)

(handwritten: $\mathscr{J} = \left(\dfrac{\partial x_i}{\partial \xi_j} \right)$ $\mathscr{J}^{\mathsf{T}} = \left(\dfrac{\partial}{\partial}\right)$)

in which, for a <u>9-noded element</u> *(handwritten: , by diff'n (3.25) w.r.t. ξ, η,)*

$$[\mathscr{J}^{\mathsf{T}}]_{2 \times 2} = \begin{bmatrix} \dfrac{\partial N}{\partial \xi} \\ \dfrac{\partial N}{\partial \eta} \end{bmatrix}_{2 \times 9} [xy]_{9 \times 2}$$ (3.29)

(handwritten: two rows into two cols.)

where, for the <u>four-noded element</u> under examination *(handwritten: , by diff'n (3.24) w.r.t. ξ,)*

$$\frac{\partial N}{\partial \xi} = \frac{1}{4}\lfloor -(1-\eta) \quad (1-\eta) \quad (1+\eta) \quad -(1+\eta)\rfloor$$ (3.30)

and similarly for the derivative with respect to η. The 2×2 matrix $\mathscr{J}$ is termed the Jacobian matrix. <u>Inversion of $\mathscr{J}$ gives the required derivatives of N with respect to x and y.</u> *(handwritten: $\left\lfloor \frac{\partial}{\partial x_i} \right\rfloor = \left\lfloor \frac{\partial}{\partial \xi_j} \right\rfloor \mathscr{J}^{-1}$)*

(handwritten margin: e.g. ϕ) The <u>functionals</u> upon which element algebraic equations are based are integrals of the derivatives $\partial N/\partial x$, $\partial N/\partial y$. To accomplish the desired integration one must replace the differential area $dA = dxdy$ with

$$dxdy = |\mathscr{J}|d\xi d\eta$$ (3.31)

where $|\mathscr{J}|$ is the determinant of $\mathscr{J}$. Also, the limits of integration become -1 and $+1$.

The foregoing procedure, although applied to the case of the four-noded element, is completely general and can be applied to representations of any order. Figure 3.11 illustrates a quadratic isoparametric element. This is constructed by writing the shape functions for the nine-noded rectangle, given earlier by Equation (3.20), in terms of the present nondimensional coordinates

(handwritten at bottom: ✻ again, author may be thinking of $F = -\nabla \phi$)

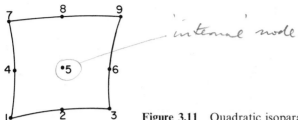

Figure 3.11   Quadratic isoparametric element

ξ and η. The Jacobian is formed, as defined by Equation (3.29), and is inverted. Integration of the functional proceeds in the manner outlined above.

The quadratic isoparametric element is perhaps the most popular of this class of formulation. It generally enables adequate representation of curved boundaries for the size of grid employed. Isoparametric elements of higher order, e.g. cubic, frequently involve ill-conditioned Jacobian matrices.

3.5   SINGULAR FUNCTIONS

Although singular conditions are of importance in electrical applications, relatively little attention has been given to finite element representations for this specific purpose. Developments motivated by needs, however, have recently brought about a fairly complete approach to the construction of singular function representations.

Numerous categorizations of singular function representation are possible.[13] In the present work we limit our attention to two categories which have the appeal of simplicity and ease of incorporation in standard finite element programs. These are based on (*a*) polynomial representations, and (*b*) isoparametric mappings.

Consider first polynomial representations. One of the earliest developments in the construction of elements with singularities[14] was accomplished simply by degenerating the bilinear rectangle (Figure 3.4) to a triangle by coalescing two adjacent vertices. This causes the values of the first derivatives of ϕ to approach infinity as the point of coalescence is approached, i.e. there is a singularity at the point of coalescence. The singularity is characterized by the term $1/r$, where r is the radius coordinate with origin at the singularity point. This early development discloses that the triangular element shape is most consistent with the representation of singularities. Consequently, in contrast with prior remarks, we discuss the triangular element in the following.

A simple polynomial field which yields a singularity of the form r^{-p} at the vertex of a triangular element, where p is any desired value, can be constructed[15] (see Figure 3.12, where the vertex 1 is the intended point of singularity). The field for this element can be described by

$$\phi = (1 - \xi^p)\phi_1 + \xi^p(1 - \eta)\phi_2 + \xi^p\eta \ \phi_3. \tag{3.32}$$

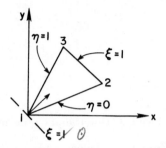

Figure 3.12 Triangular element. Singularity at point 1

In this case the (oblique) coordinates ξ and η are defined in such a way that $\xi = 1$ on the edge 2–3 and $\xi = 0$ at the vertex (point 1), and the radial edges 1–2 and 1–3 correspond to $\eta = 1$ and $\eta = 0$, respectively. It can be shown that this field gives representation of singularities of the order of r^{-p} where r is the radial coordinate in a polar coordinate system with origin at point 1.

Blackburn[16] has also proposed singular fields for the triangular element. For a six-noded element, with nondimensional coordinates defined in a manner similar to that of Figure 3.12 (i.e. with origin at the crack tip) the proposed field is

$$\phi = b_1 + b_2\xi + b_3\eta + \frac{b_4\xi + b_5\eta + b_6\xi\eta}{\sqrt{(\xi + \eta)}} \tag{3.33}$$

where ξ and η are defined as above.

This gives a linear variation of ϕ between the edges which meet at the singular vertex. Reference 17 shows that a simple modification of this field gives a quadratic variation of ϕ between such edges. Equation (3.33) forms a part of the displacement field for a three-dimensional wedge element formulated by Hellen and Blackburn;[18] an extension is given by Stern and Becker.[17]

Akin[19] has devised a scheme whereby shape functions for standard triangular and quadrilateral elements can be converted to singularity functions. As previously, we denote the standard functions as $\phi = N\phi$. Consider again the triangular element of Figure 3.12 where point 1 is the origin of coordinates. Then a special function is defined as $Q = (1 - N_1)^p$, where N_1 is the shape function for point 1 and $-p$ is the desired order of singularity. The modified shape functions $H_i (i = 1, \ldots, n)$ are:

$$H_1 = 1 - [(1 - N_1)/Q] \tag{3.34}$$

$$H_i = N_i/Q \qquad 2 \leq i \leq n. \tag{3.35}$$

For triangles the derivatives of ϕ at node 1 contain singularities of order r^{-p}.

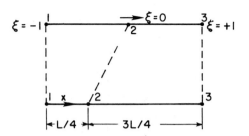

Figure 3.13 Adjustment of quadratic field
to introduce quarter-point singularity

Consider now the adaptation of isoparametric formulations to the representation of singularities. The underlying concept is most easily explained by once again considering the 3-noded one-dimensional element (Figure 3.9). The shape functions for this element were written as Equation (3.18a). Now we will specialize that expression to the case $x_2 = L/4$. Thus, with $x_2 = L/4$, $x_1 = 0$, and $x_3 = L$ (Figure 3.13),

$$x = (1 - \xi^2)\frac{L}{4} + \frac{1}{2}\xi(1 + \xi)L = \frac{L}{4}(1 + \xi)^2. \tag{3.18b}$$

In an isoparameteric formulation the transformation from physical to isoparametric coordinates is based upon the inverse of the matrix of derivatives of the former with respect to the latter. The only derivative present here is $dx/d\xi = L(1 + \xi)/2$. This gives $dx/d\xi = 0$ at $\xi = -1$, so that the transformation is singular at point 1. To investigate the consequences of this singularity we can write the field ϕ in terms of x, and differentiate. This will disclose that the singularity at point 1 is of the $1/\sqrt{r}$ type.

It can be demonstrated[20] that a variety of forms of singularity can be produced, dependent on the order (m) of the polynomial. To demonstrate this, we define both the coordinate x and the variable ϕ in polynomial form in terms of ξ, where now $0 \le \xi \le 1$.

$$x = A_0 + A_1\xi + A_2\xi^2 + \ldots + A_m\xi^m \tag{3.36}$$

$$\phi = a_0 + a_1\xi + a_2\xi^2 + \ldots + a_m\xi^m. \tag{3.37}$$

Now, the node points in the distance 0–1 can be arranged in such a way that all A_i, except A_0 and A_m, are zero. Equation (3.36) then beomces

$$x = A_m\xi^m. \tag{3.38}$$

The derivative of ϕ, which is the basis of calculation of the functional of the problem is then, after some algebra,

$$\frac{d\phi}{dx} = \frac{d\phi}{d\xi}\frac{d\xi}{dx} = \frac{\xi^{1-m}}{mA_m}\left[a_1\left(\frac{x}{A_m}\right)^{1-m} + 2a_2\left(\frac{x}{A_m}\right)^{2-m} + \ldots + ma_m\right]. \tag{3.39}$$

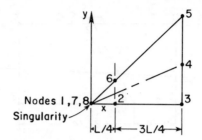

Figure 3.14 Triangular isoparametric element with $1/\sqrt{r}$ singularity

The leading term is of order $x^{(1-n)/n}$ and this gives a singularity of order $x^{-1/2}$ for $m=2$, $x^{-2/3}$ for $m=3$, and so forth. The range of singularities, as m varies from 2 to ∞, is $x^{-1/2}$ to x^{-1}.

When the above concept was first applied to a two-dimensional problem,[20-22] the generalization of the line to a quadrilateral (Figure 3.5) was adopted for planar conditions. Subsequently a 20-node hexahedron (Figure 3.8) was adopted for three-dimensional conditions. It was subsequently realized, however, that the singularity conditions prevail only along the edges for such elements, and not on an arbitrary ray emanating from the corner singularity point. The problem was resolved[24] by collapsing a pair of nodes as shown in Figure 3.14 to form a triangular element and locating the nodes as indicated. In this case the singularity prevails on all rays emanating from the vertex.

The preceding development implies that singularity elements surround the point of singularity and that only conventional elements lie beyond these. Accuracy can be improved, however, if the elements once removed from the point of singularity also include the effect of the singularity. Lynn and Ingraffea[25] show that the placement of the side-node between the quarter-point and the midpoint enables the inclusion of such 'transition' elements and a significant improvement in accuracy without additional degrees-of-freedom.

REFERENCES

1. Melosh, R. J. (1963). 'Basis of Derivation of Matrices for the Direct Stiffness Method', *J. AIAA*, **1**, 1631–1637.
2. Bazeley, G., Cheung, Y., Irons, B., and Zienkiewicz, O. (1963). 'Triangular Elements in Bending—Conforming and Nonconforming Solutions', *Proc. Conf. Matrix Methods in Struct. Mech., AFFDL TR60-80*, Oct. 1963.
3. Prenter, P. (1975). *Splines and Variational Methods*, Wiley, New York.
4. Mitchell, R. and Wait, A. (1977). *The Finite Element Method in Partial Differential Equations*, Wiley, London.
5. Gallagher, R. H. (1975), *Finite Element Analysis: Fundamentals*, Prentice-Hall, Englewood Cliffs, New Jersey.
6. Zienkiewicz, O. C. and Hinton, E. (1976). 'Reduced Integration, Function Smooth-

ing and Non-Conformity in Finite Element Analysis', *J. Franklin Inst.*, **302**, 443–461.

7. Irons, B. and Razzaque, A. (1972). 'Experience with the Patch Test'. In K. Aziz (Ed.), *Mathematical Foundations of the Finite Element Method with Applications to Partial Differential Equations*, Academic Press, New York, pp. 557–587.

8. Strang, G. and Fix, G. (1973). *An Analysis of the Finite Element Method*, Prentice-Hall, Englewood Cliffs, New Jersey.

9. Fraejis de Veubeke, B. (1974). 'Variational Principles and the Patch Test', *Int. J. Num. Meth. in Engrg.*, **8**, No. 4, 783–802.

10. Rodin, E. Y. (1976). 'A New Method of Integration over Polynomial Finite Element Boundaries', *Int. J. Num. Meth. in Engrg.*, **11**, No. 5, 1115–1124.

11. Richards, D. J. and Wexler, A. (1972). A. 'Finite Element Solutions within Curved Boundaries', *IEEE Trans., Microwave Theory and Techniques*, **MTT-20**, 650–657.

12. McLeod, R. and Mitchell, A. R. (1972). 'The Construction of Basis Functions for Curved Elements in the Finite Element Method', *J. Inst. Maths. Appl.* **10**, 382–393.

13. Gallagher, R. H. (1978) 'A Review of Finite Element Techniques in Fracture Mechanics', In A. R. Luxmoore and D. R. J. Owen (Eds), *Numerical Methods in Fracture Mechanics*, University College, Swansea. (1971).

14. Levy, N., Marcal, P. V., Ostergren, W., and Rice, J. (1971). 'Small Scale Yielding Near a Crack in Plane Strain: A Finite Element Analysis', *Int. J. of Fracture Mech.*, **7**, No. 2, 143–157.

15. Tracey, D. and Cook, T. S. (1977). 'Analysis of Power Type Singularities Using Finite Elements', *Int. J. Num. Meth. in Engrg.*, **11**, No. 8, 1225–1235.

16. Blackburn, W. S. (1973). 'Calculation of Stress Intensity Factors at Crack Tips Using Special Finite Elements', In J. R. Whiteman (Ed.), *The Mathematics of Finite Elements*, Academic Press, New York, pp. 327–326.

17. Stern, M. and Becker, E. (1978). 'A Conforming Crack Tip Element with Quadratic Variation in the Singular Fields', *Int. J. Num. Meth. in Engrg.*, **12**, No. 2.

18. Hellen, T. K. and Blackburn, W. S. (1975). 'The Calculation of Stress Intensity Factors in Two and Three Dimensions Using Finite Elements', In E. Rybicki and S. Benzley (Eds), *Computational Fracture Mechanics*, ASME Spec. Publ.

19. Akin, J. E. (1976). 'The Generation of Elements with Singularities', *Int. J. Num. Meth. in Engrg.*, **10**, No. 6, 1249–1260.

20. Hibbitt, H. D. (1977). 'Some Properties of Singular Isoparametric Elements', *Int. J. Num. Meth. Engrg.*, **11**, No. 1, 180–184.

21. Henshell, R. D. and Shaw, K. G. (1975). 'Crack Tip Elements are Unnecessary', *Int. J. Num. Meth. in Engrg.*, **9**, 495–509.

22. Barsoum, R. S. (1976). 'On the Use of Isoparametric Finite Elements in Linear Fracture Mechanics', *Int. J. Num. Meth. in Engrg.*, **10**, No. 1, 25–37.

23. Barsoum, R. S. (1976). 'A Degenerate Solid Element for Linear Fracture Analysis of Plate Bending and General Shells', *Int. J. Num. Meth. in Engrg.*, **10**, No. 3, 551–564.

24. Barsoum, R. S. (1977). 'Triangular Quarter Point Elements as Elastic and Perfectly-Plastic Crack Tip Elements', *Int. J. Num. Meth. in Engrg.*, **11**, No. 1, 85–98.

25. Lynn, P. and Ingraffea, A. R. (1978). 'Transition Elements to be Used With Quarter-Point Crack Tip Elements', *Int. J. Num. Meth. in Engrg.*, **12**. No. 6, 1031–1036.

Finite Elements in Electrical and Magnetic Field Problems
Edited by M. V. K. Chari and P. P. Silvester
© 1980, John Wiley & Sons Ltd.

Chapter 4

Software Engineering Aspects of Finite Elements

P. P. Silvester

4.1 INTRODUCTION

Software engineering is a relatively new art, having been recognized as extant since about 1970, and in retrospect traceable to perhaps 1960. It is concerned with programs and programming systems for practical use. That is to say, the task of the software engineer is to ensure that user information processing requirements are met in an effective, efficient, economic, and agreeable manner through the use of appropriately chosen computing resources.

Software engineering is not to be confused with software design; indeed the latter forms an important part, but only a part, of software engineering. The software engineer must not only design, but also implement, test, and install the necessary software. In other words, he must not only assume responsibility for choosing the correct algorithms, file structures, and languages for the task, but also for program design, program debugging, and finally for fitting the programs into the operating system and computer hardware of a specific computer installation. In addition, he is normally expected to provide user training either directly or through intermediaries, to provide documentation and maintenance support staff training, and finally to act as a consultant when difficult software maintenance problems are encountered. These various tasks are clearly interdependent: indeed it is their close interdependence which defines software engineering as a distinct art.

As in most engineering disciplines, the design function is of paramount importance. Software of good quality is, above all, usable and maintainable. These two qualities are strongly affected by the somewhat more technical requirements of *robustness*, *modularity*, and *flexibility*. For finite element software intended for use in a variety of computer installations, the property of *transportability* is additionally of importance. Within the present state of the art, there exist many techniques for achieving these goals. However, good design must take account of both the quality requirements of the application envisaged, and of the costs involved. Costs in their turn may involve capital outlays in programming or hardware acquisition, or operating expenses in the form of computer time, user training, or maintenance expense. Good software design therefore combines sound economic judgement with an appreciation of the intended applicational goals.

4.2 COMPOSITION OF PROGRAM PACKAGES

Until recent years, finite element applications software packages exhibited a clear trend toward universality; that is to say, program designers apparently hoped to create quite large-scale packages capable of solving almost any imaginable problem within a given range. More recently, there has been some tendency toward separating individual functional modules, thus creating looser program structures. Nevertheless, current packages reflect the software designer's desire for universality. As a result, applications packages tend to be fairly large. At the present state of the art, few applications packages are considered worthy of the name if they contain fewer than about 1000 executable Fortran lines, or an equivalent amount of code in other languages. Of course, much larger programs exist, the current upper limit being well beyond 10^4 lines.

In general, the majority of finite element programs written to date for electromagnetic boundary-value problems have presupposed a batch-processing environment essentially oriented to business applications—that is, an environment based on a single high-speed central processing unit with substantial amounts of rapid-access memory, and with fairly slow user response time (typically 0.3–30 hours). Such environments are normally supposed to possess as peripheral units a card reader, a line printer, essentially infinite disk or tape auxiliary storage, and sometimes a pen-and-ink plotter. Program packages have generally been designed to run in a single pass in such environments; packages which require user intervention at intermediate stages in a problem solution are somewhat more rare. It is probably fair to say that most program design has been carried out on the supposition that users do not know and do not wish to know very much about the solution methods employed. The ideal user situation envisaged, it often seems, is for the user to state his problem in clear unequivocal terms before leaving his office, returning to find the solution on his desk the following morning. In other (and perhaps less kind) words, it is assumed that users do not wish to solve problems, they wish their problems would go away and come back solved.

A strongly batch-processing and business-machine oriented attitude brings with it the consequence that the user must specify beforehand all the details of every problem, as well as all the details of the desired form of presentation of the solution. This consequence provides for the program user first an opportunity to commit many costly blunders, and secondly the sometimes unpleasant realization that just what constitutes a solution to a boundary-value problem is a highly subjective matter. A problem which may be perfectly well defined mathematically, indeed for which a mathematically unique solution is known to exist, usually does not possess a unique solution in the engineering sense. A design engineer is frequently more interested in various functionals associated with the solution than with the solution itself. Thus he may be interested in terminal impedance, dielectric loss, maximum value of electric field, not only in the field configuration itself, for the design specifications which he must meet usually present themselves as terminal quantities or overall functionals and are not stated in terms of prescribed field solutions. Most software packages must, therefore, make provision for extensive varieties of postprocessing functions, and must provide a structure of control commands enabling the user to choose which postprocessing functions he wishes to have carried out. The complexities of setting up a problem in well-defined computable form, and of stating exactly what results should be abstracted from the computed solution, are frequently underestimated by users, particularly by those well trained in classical mathematical techniques.

The art and science of creating good mathematical software are well advanced at the present time, and there exist several high-level languages (Fortran, PL/1, ...) well adapted to coding mathematical algorithms with brevity and precision. As a result, in typical program packages the number of Fortran lines devoted to carrying out the mathematical steps of actually solving a problem tends to be rather small as a fraction of the entire package.

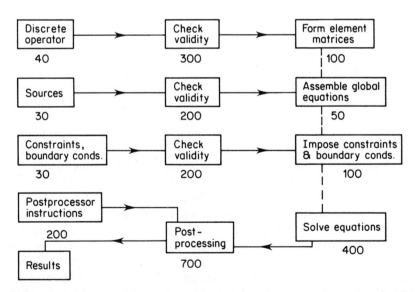

Figure 4.1 In a typical moderate-sized finite element package for electric machine analysis, numerical computation (right-hand column of operations) amounts to only 25% of total source code

By way of illustration, Figure 4.1 shows a block outline of a typical small applications package intended for solving magnetostatics problems in two dimensions. It will be noted first of all that a considerable portion of the total Fortran lines serve to validate data, and that an equally great number is required to provide for result postprocessing. It should perhaps be added that the postprocessing provided for in this package is quite modest. Indeed, a 'better' applications package for this area probably would not contain many more Fortran lines in the mathematical portion; additional package improvements would largely consist of added postprocessing functions and added input flexibility. It will be observed that only roughly one quarter of the total line count is actually involved in the mathematical operations in any real sense. The remaining three quarters, and in more complex packages an even larger portion, of code is devoted to data processing rather than numerical mathematics.

4.3   DESIGN OF FINITE ELEMENT PACKAGES

In setting out the design of a finite element package, the software engineer is faced with a large number of design choices. The correct choice in each case depends of course on the nature of the problem or problems to be solved, the prospective frequency of use, the nature of the users themselves, and obviously the computer hardware installation which it is intended to use. For the present discussion, it will be assumed that the computational tasks to be accomplished have already been decided, so that there is no question as to what to include, only how to carry it out. Generally speaking, these technical questions may be grouped under choices of mathematical method, computational method, and method of communication with the user.

In the choice of mathematical methods to be employed, two aspects predominate: the selection of a method for solving simultaneous equations, and the choice of element types. Where deterministic problems are to be solved, most existing program packages employ techniques based on Gaussian elimination. The range of such methods currently available is quite extensive. Most finite element models of realistic engineering problems involve large matrices with rather high sparsity, and it is usual to seek out techniques which take advantage of sparsity to save computer memory and time. There exist also some program packages which use iterative methods for the solution of systems of simultaneous linear equations. The choice must rely largely on considerations of robustness and problem-dependence. The convergence rate of iterative methods usually depends significantly on parameter values encountered in the problem itself, so that the total computing time involved in solving the problem depends not only on the total number of discrete variables treated but also on the numerical values encountered. Thus a selection must be made with due regard to the range of problems which the package is intended to treat. With favourable parameter values, and a range of problems sufficiently restricted to allow speedy estimation of acceleration parameters, iterative methods can prove both faster and more economic of storage than elimination methods. On the other hand, elimination techniques are very robust, hence preferable in general-purpose packages. They are somewhat more prone to round-off error problems than iterative methods, but this consideration is not a major one if iterative solution refinement can be employed. The main restriction on iterative refinement is the need to store both the original coefficient matrix itself and its triangular factors, implying either a doubling of memory requirements or the use of secondary storage devices such as disks. To a large extent, the choice of equation-solving method and the choice of elements to be used are interdependent decisions. In turn, the choice of element types is heavily dependent on the boundary shapes expected to be encountered in problems. Generally speaking, elements of high order provide much better accuracy than elements of low order, accuracy which can be traded for computing time if desired. However, their use may not be practical or appropriate in cases where very complicated many-cornered boundary shapes are encountered. The coefficient matrices resulting from the use of low order elements are also of higher sparsity than those resulting from high-order elements; if an equation-solving technique is envisaged in which high sparsity is more important than total variable count, elements of low order may sometimes be preferable.

Computational choices to be made in package design revolve primarily around the forms of internal data representation, the extent and techniques of verification of data internal consistency, the choice of input/output media, and of course the fundamental choice of batch-oriented as against interactive computing.

The internal representations chosen to describe the problem involve at least two criteria: whether repeated solutions will be required, and whether problems usually involve repetitive structures. Repeated solutions clearly imply that substantial portions of problem data should be retained in their original forms, as well as possibly in intermediate semi-processed forms; overwriting of data by the solution is to be avoided. If repeated solutions are only rarely required, there is on the other hand every good reason to conserve computer memory by overwriting. Similarly, if typical problems frequently involve repeated element groupings, or even single elements, it is often worthwhile to save memory and computing time by constructing models of subgroupings, and to use these partial models many times over to arrive at the representation of the total structure. Whether in such cases the user should identify and specify substructures, or whether their identification should be done by the program package automatically, is a very interesting problem in itself.

Batch-oriented computing usually requires rather more data verification than interactive computing, since program execution is entirely divorced from the user. Large-scale packages are rarely convenient for use if only one problem can be solved at a time, so it is usually arranged for them to return to the input stream to pick up another problem after each solution. In such looped structures it is usually inadvisable to abandon the program run if a catastrophic error is encountered; instead, provision should be made to enable the program to leaf through the remainder of the input file, looking for the next problem. Considerable annoyance and waste of time can otherwise result. Data verification and consistency checking, of course, cost computer time, and various different types of check entail different costs. Thus the designer must balance the recurrent cost of data checking against the likely cost of an occasional but potentially expensive erroneous run, the probability of any particular error being committed, and the increased programming expense of providing for ever larger numbers of checks and diagnostic messages in the program package. Syntactic tests are usually quite easy and cheap, while true tests of data deck semantics are more complex. A good finite element package might contain, for example, checks for nodes not belonging to any element (which usually leads in the assembly process to a null row and column in the coefficient matrix), tests for elements with nonpositive area or volume, and tests to ensure that boundary conditions do not mutually conflict (e.g. two distinct potentials assigned to the same node). In a batch-oriented environment it is also frequently useful to test for gross blunders: a problem description with no elements at all, a problem possessing insufficient boundary conditions or none at all, a number of boundary constraints so large as to leave no free variables, or a zero right-hand side in the assembled final equations (i.e. no sources and only homogeneous boundary conditions). Tests for intersecting or overlapping elements are possible, but relatively expensive; whether they are to be included depends, like many other decisions on data verification, on the probable level of user comprehension of the methodology. Design of data verification is naturally dependent on the design of the data formats themselves. While it is usually desirable to minimize the quantity of input data required from users, it is nevertheless often useful to demand that the user produce more than the minimally required data set, thus permitting a certain amount of redundancy in the input stream. Redundancy checks are often easily performed, and form a very strong safeguard against user error as well as against operator error in batch-oriented computing installations. (Many experienced card users are firm believers in the 'hand-span rule': any deck of cards which exceeds one human hand-span will sooner or later be dropped and shuffled by a computer operator.)

Data verification techniques and data formats themselves are strongly influenced by various problem-dependent considerations. A particularly important example arises in the solution of sequences of essentially similar problems such as a sequence of alternative trial designs differing only slightly in the materials used, geometric shapes, and the like. In such cases, it is often well to design data formats and structures in such a way that only the changed portion of the problem need be entered, it being implicitly understood that all problem parts not read in remain as previously. Of course, all consistency checks and redundancy checks must still be carried out on the altered data.

The choice of input and output media is often strongly constrained by the fact that large-scale computers, even those nominally installed for scientific purposes, are often really business machines oriented toward punched-card input and line-printer output. Indeed, truly universal portability of a program implies reliance on no peripheral devices other than these, since many commercial installations fail to provide any other peripheral devices at all. One must nevertheless keep in mind that sheer physical availability is not always an ultimate criterion. The true cost of large-scale software is very high, and such special hardware devices as a modest graphics terminal cannot be regarded as expensive even if they are not used by any other program. More significant perhaps is the ensemble of existing user habits. Users, most especially occasional users, tend to feel at home with hardware to which they are accustomed. They frequently fear hardware changes and mistrust devices they have not used previously, to the point of occasionally refusing to use program packages which they know to be of high quality. These factors tend to act together to lead program designers to adopt very conservative attitudes; indeed some entirely avoid peripheral devices other than the line-printer and the card-reader. However, it does seem needlessly wasteful of human effort and talent to produce, for example, graphical output on a line printer when modest graphics hardware is easily available. It would probably be a good idea for more program designers to make choices so as to exercise some gentle pressure on users to employ hardware suited to the purpose in view, and on computing centres to provide turn-around times more oriented to engineering problem solution than to payroll preparation.

A particularly ill-explored area which many package designers may do well to consider is the provision of interactive data editing facilities, generally at very modest cost. The point here is to enable the user to examine, and if necessary edit, input data on an interactive basis; it is not intended to force him to submit large program runs through an interactive computing system. Many input blunders can be caught by a cursory examination of all data elements prior to submitting the data for processing.

4.4 USER COMMUNICATION

Design decisions regarding communication between user and program follow to a large extent from previous decisions about the nature and scope of the program package. One may broadly distinguish problem-oriented packages and area-oriented programming systems. The former are tightly bound to a specific technological area (e.g. magnetic circuit design of transformers), while the latter attempt to cover a wide range of problems essentially similar scientifically and mathematically but quite diverse in their practical applications. Problem-oriented packages are generally much more restricted and rigid in use: indeed, frequently they are single programs with many numerical parameters accessible for modification or specification by the user. Their restricted nature makes them inflexible and difficult to modify. User command languages associated with them are usually quite rigid in syntax and format, and hence error-prone. On the other hand, they are usually very easy for users to understand, since practically all commands and options accessible to the user have direct significance in terms of the technological application envisaged. Area-oriented programming systems are usually much harder to design because the precise program flow cannot be forecast for all time. Much more sophisticated control arrangements are therefore necessary. On the other hand, the resulting modularity of individual programs makes them much easier to modify and much more flexible in use. By their very nature they tend to be considerably larger than problem-oriented packages, and lend themselves to a wider variety of problems. The price to be paid by the user is some comprehension of the package structure, for the ability to use the same programming system for a variety of problems implies that the control language employed must direct itself to the common element in all problems of the given class—usually mathematical properties of the equations to be solved rather than the technological devices or materials which the equations are intended to model.

4.5 MAJOR PROPERTIES OF APPLICATIONS PACKAGES

Software quality in general, and quality of finite element packages in particular, resides primarily in the properties of robustness, portability, flexibility, and soundness. These are properties widely understood and appreciated, but since they form part of the software art rather than science, it is difficult to endow them with precise definitions. However, this imprecision in definition in no way impedes examination of the ways in which these desirable properties may be enhanced.

4.5.1 Robustness

Software is said to be robust if it exhibits a wide tolerance for input blunders and various data inconsistencies, either resolving them in a manner sensible in the context of the problem, or else aborting the particular problem but not the entire program run; robust software should consider very few errors as fatal. Robustness also implies little or no loss of computational accuracy over a wide range of parameter variations, and lack of sensitivity to round-off error in data conversion and mathematical manipulation.

Since robustness is essentially a defensive quality, it implies foreseeing and taking defensive action against all likely disasters; the design of robust software amounts essentially to providing a gentle escape route from every foreseeable difficulty. While an exhaustive enumeration of techniques is clearly impossible, there nevertheless exist some widely applicable methods for enhancing this quality.

Protection against input blunders and data errors is gained by good input data design and by comprehensive consistency checking. Well-designed input arrangements are first and foremost fault-tolerant. Users should be allowed considerable syntactic freedom; for example, numerical input should not be critically dependent on positioning in particular card columns, and independent segments of a data set should not necessarily be read in a fixed sequence. Secondly, well-designed input data will keep data sets small, but will include sufficient redundancy to allow for consistency checking. Thirdly, the data items required should be as nearly problem-oriented or area-oriented as is reasonably feasible, for users will generally spot blunders easily in a data set if its form is not far removed from their daily experience. Good design also implies numerous checks and tests, independent of each other and verifying independent aspects of the data. Equally important is the action taken in response to errors. If data formats and structures are well designed, it is unlikely that a built-in error option will guess correctly which of two conflicting data portions is correct. Thus it is usually better to abandon the particular problem in hand rather than to waste computer time in continuing with a data set which may or may not be correct. Instructions or option specifications which might potentially result in catastrophic action (ranging from file destruction to expenditure of large quantities of computer time) should involve instruction redundancy. Instruction redundancy may be achieved by requiring double commands. For example, instead of a single command ERASE FILE 6 in a batch-oriented environment, two commands on separate input cards might be required: READY TO ERASE FILE 6 and ERASE FILE 6. These two instructions involve two forms of redundancy: first, they ensure that the command ERASE is really what the user intended, and secondly, they permit a cross-check to make absolutely certain that the user really meant file number 6. In iterative computing environments instruction redundancy is usually, perhaps not very imaginatively, achieved by issuing a prompt for a repetition of the instruction. Thus, ERASE FILE 6 might result in the reply DO YOU WANT ME TO ERASE FILE 6? with action taken only after receiving a YES.

Robustness in the mathematical and arithmetical sense is achieved above all by the choice of a stable and well-functioning algorithm. Numerical stability and round-off-proofing can be enhanced considerably, however, by wise use of double precision computation. For example, inner products of vectors (equivalently, matrix–vector multiplications) are best accumulated in double precision. The extra memory requirements here are clearly negligible, since the number of double precision storage locations required is either three or one, depending on the nature of the computer system employed. Similarly, finite element matrices are often best computed in double precision, and rounded to single precision only when accumulated into the overall coefficient matrix. Solutions of simultaneous equations can be greatly improved by iterative refinement; again, one double precision storage location is required, as well as a secondary storage peripheral device on which a copy of the original coefficient matrix can be saved.

4.5.2 Portability

Program portability from one computing installation to another is in theory usually achieved by employing a well defined standard programming language (e.g. Fortran IV, Algol 68, PL/1) and adhering strictly to the language definition. In fact, this goal is almost impossible to reach, and in any case it is not at all clear that it is desirable to reach it.

To begin, programming is often more efficient in nonstandard dialects than in the standard language. For example, nonstandard dialects of Fortran permit multiple subroutine entries and exits, core-to-core I/O, and implicit type declarations. On a more mundane level, nonstandard dialects frequently support very general forms of integer subscripts and nested subscripts. Some dialects also allow word length changes, which may well result in significant core savings where integer pointers, such as node numbers, are known to remain always within a relatively restricted range.

Although formally most high-level languages are word-length independent, virtually all mathematical operations are sensitive to word length. Thus results achieved on one computer are not necessarily reproducible on another. Furthermore, most alphanumeric input–output operations are word-length dependent, since the number of alphabetic characters per machine word is not always the same. For example, object-time format specifications to Fortran programs are only decoded at execution time, and are therefore entirely dependent on machine word length. If the word length is reduced, object-time format specifications lose some of the characters transmitted by the input device, with unpredictable results in decoding. Should the word length be increased in the new installation, additional blank characters are inserted, with equally unpredictable results.

Efficient use of available memory is frequently achieved by overlaying program segments. The extent of overlaying permissible, and the instructions for achieving overlays, usually depend on the operating system and the computer employed. Thus any programming system employing overlays cannot be totally portable, a certain amount of redesign being necessary when changing computers.

Most finite element packages are intended for repeated use. It is of considerable interest therefore to code in as efficient a manner as possible those operations which are performed over and over again, for example the scalar product of two vectors. Best performance is frequently achieved by coding such operations in assembler language. Frequently-called subroutines can often be speeded up by rewriting subroutine linkages in assembler language. Portability is then retained only for other installations using a similar type of computer.

Finally, but very importantly, it should be stressed that writing programs in truly standard Fortran or Algol is virtually impossible. There can only be one acceptable proof of a program having been written in the standard language: its successful compilation by a compiler oriented to precisely the defined language standard, and subsequent successful execution. Such compilers, however, are very rare: indeed, there exist many major computer installations which offer the user a broad spectrum of compilers, each prepared to accept a slightly different dialect but none adhering to the defined standard. It should be emphasized that syntax analysis alone is not sufficient to determine whether a program will run successfully. Subroutine calls as well as Fortran COMMON and EQUIVALENCE specifications can cause various side effects that differ between computing installations, but which do not show up in syntax analysis.

Portability thus cannot in general be achieved through adherence to standard language definition, and it may well be undesirable to try. However, a high level of portability for the program package may be achieved by providing documentation, as detailed as possible, on all nonstandard language features used in programming, and on the precise objectives achieved by any assembler language code segments. In most circumstances recoding to avoid some particular language feature is easily accomplished—provided one knows exactly where and what needs to be recoded!

4.5.3 Flexibility

Programs should be considered flexible if they can be altered to accommodate various new features without extensive re-programming. Flexible programs are easy to maintain in the long run, and flexibility frequently also contributes to portability. The essential key to flexibility is program modularity. That is, programs are usually fairly flexible if they consist of interlinked modules with loose and well-defined linkages.

A level of flexibility which permits easy program maintenance is usually achieved if each module constitutes a logical program unit, sufficiently clear-cut in its purpose and sufficiently loosely linked to other modules to allow individual testing in a simulated environment. It should adhere to the same data formats and the same general structural features as other modules in the same program package. Insofar as is practical, it should adhere to the ideals, if not the strict definitions, of structured programming. Communication links with other modules should be clear-cut: there should be no side effects. For the Fortran programmer, this usually means avoiding large numbers of labelled COMMON memory allocation statements. Where modules form subroutines, they should normally not contain fixed array dimensions or other fixed limits. Finally, each module should be thoroughly and individually documented both as to its internal functioning and its external objectives.

4.5.4   Soundness

A program can only be considered sound if it has been thoroughly tested. There are two levels of testing which should be considered essential: module testing and package testing.

Each module should be thoroughly tested in a simulated environment in three distinct ways. It should first be logically checked out, to ensure that all possible logic flow paths actually have the intended effect. It should be tested mathematically for programming blunder as well as for numerical error propagation. Finally, it should be tested against known data so as to assess its numerical soundness.

Package testing should primarily direct itself to the external appearance of the entire package. Thus, all reasonably possible forms and combinations of input data which are considered legal should be tried out in the testing. The program package should, where applicable, be checked out in direct competition with other programs capable of solving the same mathematical problem, preferably using the same type of model; finite element programs should be checked against other finite element programs. Finally, checks against known physical data ought to be performed as a final verifier.

Complete documentation should be kept on file of all validation tests, including a few printouts.

Every program should be tested at the package level not only by its designers, but by numerous novice users, for its ergonomic soundness. Default options which seemed reasonable to the designer may not seem so to novice users, and data sequences which appeared convenient to the programmer may not seem so to the user. Novice users are generally much more sensitive to problems of this sort than old hands, and are therefore to be preferred as guinea pigs. In particular, any changes introduced in response to user criticism should be retested by at least a few users not previously acquainted with the old version.

4.6   DOCUMENTATION AND MAINTENANCE

Software needs not only to be written; it needs also to be documented so others can use it, and it needs to be maintained over fairly long periods of time. The need for documentation is self-evident to most engineers, whether their past experience be software or hardware. The need for continuous maintenance, on the other hand, appears somehow wrong to persons with a hardware background, for it seems that programs should not wear out, and therefore should need no maintenance. A few words on the need for maintenance might therefore be in order.

4.6.1 Maintenance of software

There are four basic reasons why finite element software packages need continuous long-term maintenance: flaws in the original program, changes in the program environment, changes in the problems to be solved, and changes in the finite element method.

No matter how carefully constructed a package might be, and no matter how carefully it is tested, some residual program errors will persist. Errors may also persist in its installation into the operating environment of the host computer. As users gain operating experience with any program package, they gradually increase their demands. Program bugs can remain hidden for periods that may stretch into years, simply because the combination of circumstances which leads to trouble had never occurred before.

Changes in program operating environment tend to occur far more frequently than most users are aware. Every computing centre hardware change causes some changes in file access procedures, I/O channel capacities, and possibly in the availability of actual physical devices. Most computing centres attempt to make such changes upward-compatible, so as not to inconvenience users too much. However, the new procedures or peripheral devices may well be potentially very useful, making it desirable to modify software even though modification is not necessary for survival. Hardware changes in the central processor, on the other hand, may change numerical precision (e.g. because of altered word-length), or may change the amount of available memory. It must be remembered that an applications package does not operate merely in a hardware environment, but in an environment composed of hardware and system software. The system software of any major computing centre undergoes continuous maintenance and alteration, which in turn will impinge on user software. Again, most system managers attempt to make alterations upward-compatible. But since they wish to phase out old system software in time, managers generally encourage users to modify their software to suit the system, by various pressure mechanisms ranging from service deterioration for users of the 'old' system, to gradually increasing special charges. Finally, it must be remembered that all program packages ultimately operate in an economic environment, and that a comparatively minor alteration in a computer installation's charging rates and policies can hit a given user very hard. System managers usually strive to keep all operating hardware fully occupied, and to this end frequently adjust charging algorithms so as to encourage users to alter the hardware requirements of their jobs. For example, idle tape drives can be put to work by increasing the fee levied for card handlings, so as to encourage users to employ magnetic tape files. Major reprogramming may be necessitated by such pricing shifts.

Applications packages also require maintenance in response to changes in problem definition and scope. As user confidence and abilities grow, new problems may arise which are within the mathematical scope of the package, but for which the input data description language is not well suited. In addition, data consistency checks are usually problem-dependent, and checks valid for the old type of problem may not be suitable for the new one. Default options provided at the input or output may turn out to be the wrong way around for the new problem, or better data formats (e.g. to suppress repetition) may be desirable. Even if the problem itself is not changed, new postprocessing functions may well be required. If other software is to postprocess the output, reformatting of output data may become necessary. Finally, changes in internal procedures may be desirable. For instance, it may be of interest to produce solutions to sequences of closely related problems without the necessity for re-entering full problem data each time.

An important aspect of maintenance is response to changes in engineering science. These are generally of two kinds: changes in essentially mathematical processing techniques, and changes in the finite element method. Advances belonging to the first category may include improved methods for handling sparse matrices and improved methods for numerical integration around singularities. The second category might include new elements which duplicate the function of already existing ones, but perform better; or entirely new elements which permit altogether new problems to be solved. It may offhand appear that all such changes may have their impact minimized by strong program modularity, e.g. by accessing each element by a separate subroutine call. However, while modularity certainly is desirable, it does not solve all problems. For example, improved methods of sparse matrix handling may require entirely altered forms of storage in memory, so that insertion of an improved equation-solving routine cannot be achieved by simply substituting one subrouting package for another. New elements to do an old job may be directly substitutable; but new elements to do new jobs will almost certainly entail substantial alterations because they alter the program environment, as discussed above.

While there will be occasions on which major reprogramming of an applications package is undertaken, the majority of maintenance operations only require changing a few lines of source code. Such changes can be effected quickly and inexpensively if the software package is well structured from the outset, and if it is well documented.

4.6.2 Documentation

An axiom of the software art is that nearly all programs are badly documented by their authors. This situation may be unavoidable, for those human qualities which make good programmers do not make good technical writers. It is not clear whether this assertion is strictly true; but good programmers will probably always produce bad documentation of their own programs, just as great performing artists rarely make good music critics. In other words, software documentation should not be produced exclusively by the program author, but should have strong participation by at least one program user.

Program documentation for any given package should in any case not consist of a single document. Indeed, documentation on three entirely distinct levels should ideally be provided: for the software maintenance engineer, for the program user, and for the analyst.

Maintenance documentation should be as complete as at all possible, a requirement which usually also makes it difficult to read. Maintenance documentation should clearly state on what hardware the program was developed, under what operating system, and with what compilers. It should state clearly what nonstandard dialect features of the programming language were employed. It should make entirely clear what memory management instructions (e.g. Fortran COMMON, EQUIVALENCE) are employed, and what side effects they may cause. Finite element programs typically contain many large arrays, some of which share memory; furthermore, many of them are dimensioned interdependently. The reasons for the specific array dimensions chosen, and instructions for their alteration, should appear. Full symbol tables, with verbal definitions of variable names, are useful. Maintenance documentation should always be accompanied by extensively commented listings of the source programs, but it must not be imagined that provision of commented listings is an adequate substitute for maintenance documentation.

The primary purpose of user documentation should be a clear statement of how programs interface with the user, with other programs, and with the expected hardware configuration. Input data formats should be clearly specified, along with all user-accessible options and variables. Default options should be listed, and methods for changing default options should be clearly stated, if any exist. Output variables names as well as input and output file structures should be clearly specified. Limitations on input and output data, whether arising from array dimensioning internally or form word-length-dependent precision limitations, should be clearly stated. However, user documentation ought not contain most of the internal details normally expected in maintenance documents. Every user's manual should contain an extensive set of sample input data, together with the output produced.

Descriptive documentation should accompany any serious program package. It should permit the user to understand the methods employed, and allow the prospective user to evaluate the likely usefulness of the package for his purposes. The mathematical and computational methods employed should be discussed here, as should their limitations. The types or classes of problems for which the package is believed particularly well suited should be evident. Data checking and other data processing algorithms should be outlined. Nonstandard language features, as well as hardware and system software requirements, should be indicated in sufficient detail for evaluation, but not so extensively as to confuse the casual reader.

4.7   FUTURE REQUIREMENTS

During the late nineteen-seventies, the world of computer hardware is undergoing a profound revolution, with complexity growths and price decreases around 30% annually. Quite clearly, the existence of massive amounts of cheap hardware will strongly influence software development. However, it is to be expected that the growth of software complexity will not parallel the hardware explosion.

At the present time, applications software packages are considered large if they contain about 5000 Fortran lines or more. It appears that single packages beyond 10 000–20 000 Fortran lines form single units very costly and very difficult to maintain. It is not likely that single packages very much larger than this will ever be fruitful. The hardware revolution which may be expected to continue into the middle eighties will probably simplify software a little, by allowing increased use of automated proving of program correctness. However, on balance it seems that hardware growth will make software maintenance more difficult by subjecting software packages to a rapid and continuous alteration of their hardware environment.

To satisfy finite element user needs of the eighties, programs involving the equivalent of 10^4–10^5 executable Fortran lines will probably be needed. Unless some as yet undefined fundamental breakthrough occurs in the software engineering art, the maintenance of single packages, or groups of tightly linked packages, will be virtually impossible. Keeping in mind the very cheap memory prices now in prospect, it is likely that future software package development will incline towards the construction of special-purpose programming systems, rather than programs. The distinction here is between tightly interlinked subroutines or coroutines on the one hand, and free-standing program modules loosely linked by common system commands and file structures on the other. In the latter case, little or no communication between programs will or should occur through shared memory; instead, each program module may be regarded as a processing device that operates on given files. In other words, the work of the practising finite element analyst is likely to resemble the work of the commercial data processing specialist to a greater degree. It will become more file-oriented and less procedure-oriented. The individual processing modules may be essentially hardware-independent, but their use will involve programming in special-purpose job-control languages, which will be tailored to the local hardware configuration.

To summarize, the software engineering requirements of future applications programs are likely to be satisfied only by highly modular programs capable of operating on common files even at the expense of memory or processing speed. The high degree of modularity is required first of all to make programs maintainable, and secondly to allow users much greater flexibility than they are presently accustomed to. Naturally, modular programs must be linked together by special-purpose programming or operating systems. They must be supported by purpose-oriented documentation, equally modular and equally standardized. By adopting this approach, finite element software engineers are likely not only to survive the hardware revolution of the eighties, but to turn it to immense profit.

Finite Elements in Electrical and Magnetic Field Problems
Edited by M. V. K. Chari and P. P. Silvester
© 1980, John Wiley & Sons Ltd.

Chapter 5

Finite Element Solution of Magnetic and Electric Field Problems in Electrical Machines and Devices

M. V. K. Chari

5.1 INTRODUCTION

The past two decades have witnessed phenomenal growth in electrical power systems and sizes of electrical plant such as transformers, salient pole alternators, direct current machines, accelerator magnets, turbine-generators, and a host of other electrical devices. An accurate prediction of their performance has, therefore, become increasingly important to meet stringent specifications, to effect economy in design, and to ensure reliability of operation. Some of the performance indicators that machine designers and power systems engineers are vitally concerned with are the excitation requirements under open circuit, short circuit, and full load conditions, sequence reactances, transient characteristics, short circuit forces, iron and stray load losses, endfield and eddy current effects in the case of a.c. machines, load regulation and commutation characteristics in d.c. machines, insulation strength, and dielectric withstand of surge voltages of various parts of electrical machinery.

In order to evaluate these, it is imperative that the electric and magnetic fields under various operating conditions are predicted accurately. In view of the inadequacies of classical analytical methods and analogue techniques, the need for numerical solutions was recognized even in the early stages of the design art. Nevertheless, only with the advent of large scale digital computers could such methods be developed and extensively used for solving the field distribution in electrical machines with great detail and precision.

All of the numerical methods in present day use fall under three principal headings, namely: (i) Divided difference schemes, (ii) Integral equation techniques, and (iii) Variational formulations.

The variational method consists of formulating the partial differential equations of the field problem in terms of a variational expression called the energy functional. In most engineering applications this expression can be identified with stored energy in the system. In general, the Euler equation of this functional will yield the original differential equation. The minimization of the energy functional is implemented in the finite element method, whereby the potential function approximating the true solution is defined in discretized subregions of the field region.

In this chapter various applications of the finite element technique for linear and nonlinear electric and magnetic field problems are presented. Some of the areas surveyed are rotating electrical machinery cross-sections and the end-region, transformers, diffusion problems for magneto-telluric survey, and eddy-current analysis in conducting media and electrostatic applications.

The finite element method was first used for analysing saturation effects in acclerator magnets by Winslow,[31] who used a restricted functional formulation based on the assumption of fixed reciprocal permeability. The set of trial functions was defined in a discretized region consisting of finite regular meshes of triangular elements of variable geometry, but fixed topology. This restriction coupled with the slow convergence of the iteration method used did not result in any computational gain over the divided difference approach. Variants of this method for two-dimensional magnet fields and axisymmetric problems have been described by Colonias.[12]

5.2 NONLINEAR MAGNETIC FIELD ANALYSIS OF ELECTRIC MACHINE CROSS-SECTIONS BY THE FINITE ELEMENT METHOD

A general nonlinear variational formulation of the electromagnetic field problem in electric machines was first presented by Silvester and Chari,[23] and later by Anderson, Silvester, Cabayan, and Browne,[26] Glowinski and Marrocco,[17] Kreisinger,[20] and Brandl, Reichert, and Vogt.[5] In this analysis, some of the shortcomings of earlier methods are overcome resulting in economy and efficiency of computation of the field problem. The method will be briefly described hereunder for the sake of completeness.

The nonlinear partial differential equation for a magnetostatic problem in a two-dimensional cartesian system is given by

$$\frac{\partial}{\partial x}\left(v\frac{\partial \mathbf{A}}{\partial x}\right)+\frac{\partial}{\partial y}\left(v\frac{\partial \mathbf{A}}{\partial y}\right)=-\mathbf{J} \tag{5.1}$$

The field problem described by Equation (5.1) can be expressed in variational terms as an energy functional, the minimization of which yields the required solution:

$$\mathscr{F}=\iint\left[\int_0^B vb\,db - \mathbf{J}\cdot\mathbf{A}\right]dxdy - \oint_s \mathbf{A}\cdot\frac{\partial \mathbf{A}}{\partial n}ds \tag{5.2}$$

where

$$b = \sqrt{\left\{\left(\frac{\partial \mathbf{A}}{\partial x}\right)^2 + \left(\frac{\partial \mathbf{A}^2}{\partial y}\right)\right\}}. \tag{5.3}$$

By setting the closed line integral of Equation (5.2) to zero, natural boundary conditions (namely homogeneous Neumann boundary conditions) are implicitly satisfied by the energy formulation.

First-order triangular finite elements of unrestricted geometry, topology, and containing material inhomogeneities are used for discretizing the field region as shown in Figure 5.1. The potential solution is defined in terms of shape functions of the triangular geometry and its nodal values of potential. Thus

$$\mathbf{A} = \sum_{k=i}^{m} \zeta_k \mathbf{A}_k. \tag{5.4}$$

The minimization of the functional is performed with respect to each of the nodal potentials as

$$\frac{\partial \mathscr{F}}{\partial \mathbf{A}_k}\bigg|_{k=i,j,m} = \sum_{k=il=i}^{m} \sum^{m} (b_k b_l + C_k C_l)\mathbf{A}_k - \mathbf{J}\left(\frac{\partial \mathbf{A}}{\partial \mathbf{A}_k}\right) = \mathbf{0}. \tag{5.5}$$

When the minimization described by Equation (5.5) is carried out for all the triangles of the field region, the following matrix equation is obtained, solving which the unknown vector potential is determined.

$$[S].[\mathbf{A}] = [T]. \tag{5.6}$$

The matrix $[S]$ is nonlinear, symmetric, sparse, and band structured.

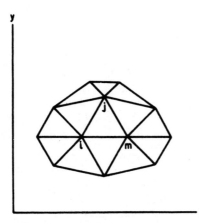

Figure 5.1 Discretization of the field region by triangular finite elements

Therefore, a special single indexing routine is employed to store the coefficient matrix with utmost economy. Since Equation (5.6) is nonlinear, it is first quasi-linearized by a Newton–Raphson algorithm and the resulting matrix equation is solved directly. In each Newton–Raphson iteration, the reluctivity is updated with respect to the *B–H* characteristic of the material. The Newton–Raphson algorithm and the solution of the linearized equations are as follows.[8]

The kth iterate of the vector potential yields the $(k+1)$st iterate, following the relation

$$\mathbf{A}^{k+1} = \mathbf{A}^k - [\mathscr{J}]^{-1}[\mathbf{S}\mathbf{A}^k - \mathbf{T}]$$

where $[\mathscr{J}]$ is the Jacobian matrix of the partial derivatives of the iteration function $[\mathbf{S}\mathbf{A}^k - \mathbf{T}]$.

The set of linearized equations resulting from Equation (5.7) reduces to a matrix equation, and the coefficient matrix is symmetric, sparse, and band-structured. A gaussian elimination technique is employed which yields a stable and accurate solution. The **B–H** characteristic may be modelled by a piecewise linear interpolation method with a large number of intercepts to ensure slope continuity.

The magnetic field plot obtained for a turbine generator on no-load[9] is shown in Figure 5.2. Although relatively few triangular finite elements were used in the analysis, the flux-plot indicates the general form of the vector potential solution obtained.

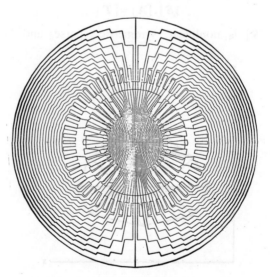

Figure 5.2 Flux plot of a turbine generator on no-load

Silvester, Cabayan, and Browne,[26] applied the Newton–Raphson process to the functional formulation instead of the residual vector $[SA^k - T]$. By this method, computational advantages are said to obtain over earlier methods. The new formulation has the attractive feature that a theoretical foundation is laid for high-order elements. The Newton–Raphson algorithm applied to the functional minimization is given by

$$A^{k+1} = A^k - \left[\frac{\partial^2 \mathscr{F}}{\partial A_i \partial A_y} \right]^{-1} \left[\frac{\partial \mathscr{F}}{\partial A_i} \right] \tag{5.8}$$

where

$$\mathscr{F} = \int_u \left[\int_0^B vbdb - \mathbf{J} \cdot \mathbf{A} \right] du. \tag{5.9}$$

The modelling of the **B–H** characteristic has been implemented by piecewise cubic spline interpolation, which has general application to material characteristics in different regions, and is more accurate than the simple piecewise linear interpolation technique. Another important contribution made by the authors to machine field analysis is the use of the Cuthill–McKee node-numbering algorithm. This procedure makes effective use of the sparsity of the coefficient matrix and consequently should result in reduced cost of computation.

Silvester and Rafinejad[28] described a similar method to the above employing higher order curvilinear finite elements.

5.2.1 Modification to the energy functional

Brauer[6] proposed a modification to the energy functional for the term relating to the forcing function, so that

$$\mathscr{F} = \int_V \left[\int_0^B \mathbf{H} db - \int_0^A \mathbf{J} d\mathbf{A} \right] dV \tag{5.10}$$

where $\mathbf{H} = v\mathbf{B}$ and $\mathbf{A}$ is the vector potential. It is stated that for magnetic conductors and other magnetic media, formulation of the functional described by Equation (5.10) yields more accurate results than that defined by Equation (5.2).

5.2.2 Magnetic field solution by iterative methods

Andersen[1] described an iterative solution of the nonlinear equations resulting from the finite element discretization and functional minimization. The reluctivities in the various triangular subregions are updated, satisfying the (v)

relation

$$v_{new} = v_{old} + f_v(v - v_{old}) \tag{5.11}$$

where f_v is a deceleration factor usually set at a value of 0.1.

The set of linearized equations in each pass are solved iteratively with an acceleration factor of 1.9 for the vector potentials. In order to avoid numerical instability, it is suggested that the conductors are initially assigned a permeability greater than unity and gradually reduced to the free space value. By this method of formulating the field problem, it is stated that an efficient and economical solution is obtained. An added feature of this work is that an automatic grid generation routine is incorporation in the program.

Glowinski and Marrocco[17] applied the finite element method for the nonlinear magnetic field analysis of a salient pole alternator. The nonlinear functional is minimized and quasi-linearized by the Newton–Raphson algorithm, and the resulting set of equations is solved by a successive point over-relaxation technique. Convergence proofs for the finite element approximation and iterative algorithms are presented by the authors, as also by Unterweger.[29]

5.2.3 Magnetic flux distribution on load

When a rotating electrical generator is delivering power to a load, the axes of symmetry of the magnetic field depart a good deal from the polar and

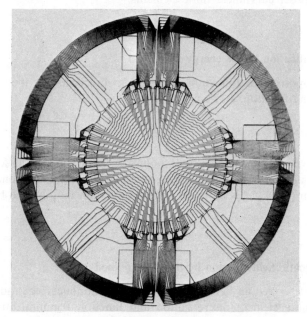

Figure 5.3   Magnetic flux distribution in a d.c.
generator on load

interpolar axes. Under these conditions, the vector potential along the pole axis at any point is of equal magnitude to that of the corresponding point a pole pitch away but of opposite sign. This is commonly known as the 'periodicity condition', which can be considered as an additional boundary condition. The coefficient matrix and forcing function of Equation (5.6) are, therefore, modified so that the off-diagonal terms corresponding to the points where the periodicity condition applies are reversed in sign. This process is implemented by means of a connection matrix.[8] Figure 5.3 illustrates the shift in the magnetic axis on load for a d.c. generator.

Brandl, Reichert, and Vogt[5] applied the finite element method for analysing the field distribution in a turbo-generator on steady state load for various power factors. It is shown that the voltage, real and reactive power, impose subsidiary boundary conditions which have also to be satisfied by the calculated field distribution. With sinusoidal time variation of flux, the induced voltage is linearly related to flux linkages and hence also to the vector potential. The flux linkages in the direct and quadrature axes are calculated in terms of the vector potential. These linkages must, however, yield the direct and quadrature axes voltages accurately. In general this is a two-step iterative process, though the authors have adopted an ingenious alternative scheme employing the Newton–Raphson technique to solve a set of simultaneous equations, whose starting values are given by

$$[S][A] - [T] = [R_1] \qquad (5.12)$$

$$[C_1][A] - E \sin \theta = R_2 \qquad (5.13)$$

$$[C_2][A] - E \cos \theta = R_3. \qquad (5.14)$$

The Newton–Raphson algorithm for the above set of equations is given as

$$\frac{\partial S}{\partial A} \Delta A - \frac{\partial T}{\partial I_p} \Delta I_p - \frac{\partial T}{\partial \theta} \Delta \theta + [R_1] = 0 \qquad (5.15)$$

$$\frac{\partial C_1}{\partial A} \Delta A - E \cos \theta \, \Delta \theta + R_2 = 0 \qquad (5.16)$$

$$\frac{\partial C_2}{\partial A} \Delta A + E \sin \theta \, \Delta \theta + R_3 = 0 \qquad (5.17)$$

where C_1, C_2 are row vectors, defined by $C_1 A = E \sin \theta$, $C_2 A = E \cos \theta$; θ is the angle between the pole axis and the rotor mmf; E is the internal voltage of the generator; I_p is the field current; and $[R_1]$, R_2, and R_3 are residuals.

The resulting set of matrix equations is band structured for the most part, but nonzero terms are shown to arise in the last two rows and columns, which appear to offset the advantages of a banded matrix in respect of computer storage and execution time.

5.2.4 Anisotropic formulation

Wexler[30] first proposed an anisotropic formulation of the two-dimensional Poisson equation for application to a reluctance machine. The magnetic permeability and field intensity are shown to be related through the permeability tensor

$$\mathbf{\mu}' = \begin{bmatrix} \mu_x', 0 \\ 0, \mu_y' \end{bmatrix} \tag{5.18}$$

where μ_x' and μ_y' are the permeabilities along the local principal axes x' and y'. A global coordinate system x, y is defined which is inclined to the local axes at the angle θ. Field vectors, expressed in these systems, are related by

$$\mathbf{B}' = \mathbf{T}\mathbf{B} \tag{5.19}$$

$$\mathbf{H}' = \mathbf{T}\mathbf{H} \tag{5.20}$$

where $\mathbf{T}$ is the unitary tensor

$$\begin{bmatrix} \cos \theta & \sin \theta \\ -\sin \theta & \cos \theta \end{bmatrix} \tag{5.21}$$

The above method provides for the existence of a source term in the scalar potential formulation. In the partitioning of the region, the deleted space is replaced by a virtual source distribution along the interface. The following relations then hold.

$$\nabla.\mathbf{B} = p \tag{5.22}$$

$$\mathbf{H} = -\nabla\phi. \tag{5.23}$$

Substituting for $\mathbf{B}$ and $\mathbf{\mu}$ appropriately from Equations (6) and (7) of reference 30, one obtains the relation

$$\frac{\partial}{\partial x}\left(a\frac{\partial\phi}{\partial x} + b\frac{\partial\phi}{\partial y} \right) + \frac{\partial}{\partial y}\left(b\frac{\partial\phi}{\partial x} + c\frac{\partial\phi}{\partial y} \right) = -p \tag{5.24}$$

where

$$a = \mu_x' \cos^2\theta + \mu_y' \sin^2\theta \tag{5.25}$$

$$b = (\mu_x' - \mu_y') \sin \theta \cos \theta \tag{5.26}$$

$$c = \mu_x' \sin^2\theta + \mu_y' \cos^2\theta. \tag{5.27}$$

The functional formulation for the anisotropic scalar Poisson equation

(5.24) is obtained as

$$F = \int_R \left\{ \left(a\frac{\partial u}{\partial x} + b\frac{\partial u}{\partial y} \right)\frac{\partial u}{\partial x} + \left(b\frac{\partial u}{\partial x} + c\frac{\partial u}{\partial y} \right)\frac{\partial u}{\partial y} - 2up \right\} dR \qquad (5.28)$$

where u is a scalar function of position.

Functional minimization and equation solutions are obtained by the finite element algorithm.

5.3   ANALYSIS OF END-REGION FIELDS OF ROTATING ELECTRICAL MACHINES

A magnetic field analysis in cartesian coordinates for the end cross-section of a turbine-generator is shown in Figure 5.4 for no-load conditions.[10] The rotor end-winding current and vector potentials were assumed to be unidirectional and orthogonal to the cross-section of the region.

The analysis includes eddy currents in the conducting screen which protects the clamping flange. Initially a magnetostatic analysis is carried out, from the solution of which the average vector potential on the screen is calculated. This

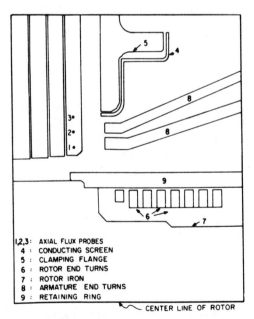

1,2,3: AXIAL FLUX PROBES
4 : CONDUCTING SCREEN
5 : CLAMPING FLANGE
6 : ROTOR END TURNS
7 : ROTOR IRON
8 : ARMATURE END TURNS
9 : RETAINING RING

CENTER LINE OF ROTOR

Figure 5.4 Section through end-region of turbine-generator. 1, 2, 3, axial flux probes: 4, conducting screen; 5, clamping flange; 6, rotor end turns; 7, rotor iron; 8, armature end turns; 9, retaining ring

average potential subtracted from the initial screen potential, and the net vector potential obtained is prescribed on the conducting screen as a boundary condition for a second field solution. The expressions relating to the above procedure are given below.

$$\mathbf{A}_{avg} = \frac{1}{N} \sum \mathbf{A}_i \tag{5.29}$$

$$\mathbf{J} = -j\frac{\omega}{\rho}\mathbf{A}. \tag{5.30}$$

The total eddy current in the screen is then given as

$$\mathbf{I}_e = \frac{-j\omega}{\rho} \sum_i^N (\mathbf{A}\Delta S). \tag{5.31}$$

Also

$$\mathbf{I}_e = \frac{-j\omega}{\rho} \mathbf{A}_{avg} S \tag{5.31a}$$

where S is the cross-sectional area, ω is the angular frequency, and ρ is the resistivity of the conducting screen.

From equations (5.31) and (5.31a), we have

$$\sum_i^N (\mathbf{A} - \mathbf{A}_{avg})\Delta S = \mathbf{0}. \tag{5.32}$$

Equation (5.32) suggests that the potentials on the conducting screen must be replaced by $(\mathbf{A} - \mathbf{A}_{avg})$ in order to satisfy the special boundary condition that the eddy currents in the screen must sum to zero.

The resulting flux distribution is shown in Figure 5.5. A good correlation with factory tests was obtained by this method in the region of interest.

Okuda, Kawamura, and Nishi[21] described a vector potential solution of the end-field problem in cylindrical polar coordinates. The field problem is formulated as a diffusion equation employing three components of vector potential and one scalar potential. In this analysis, the absolute vector potential $\mathbf{A}'$ is defined in terms of the floating vector potential $\mathbf{A}$ and a scalar potential function $\emptyset$ as

$$\mathbf{A}' = \mathbf{A} + \nabla\emptyset. \tag{5.33}$$

The energy functional for the field problem is given as

$$\mathscr{F} = \int\int\int_v \left[\frac{B_t^2}{2\mu} - \mathbf{J}\mathbf{A} + j\frac{\omega\sigma}{2}(\mathbf{A} + \nabla\phi)^2 \right] r\,dr\,d\theta\,dz \tag{5.34}$$

Figure 5.5   Magnetic field distribution on no-load with eddy current shielding by the conducting screen

where the vector potential **A** has three components in the r, θ, and z directions respectively.

First order triangular finite elements are employed and the functional minimization is carried out with respect to each of the nodal components of the vector potential and one scalar potential. This results in a (12×12) skew symmetric matrix for each triangle. The respective matrices for the various triangular discretized regions are adjoined in the usual manner and the equations are solved by complex arithmetic. Flux plots for no-load and three-phase short circuit conditions are presented with others and comparison of local flux distribution with search coil measurements is illustrated.

Howe and Hammond[18] applied the finite element analysis, which included eddy currents, to end field problems of simplified geometry in a two-dimensional cartesian system.

5.4 TRANSFORMER FIELD ANALYSIS BY FINITE ELEMENTS

Early work on finite element analysis of transformers was presented by Silvester and Chari.[23] A quarter section of a single phase transformer was considered with nonlinear magnetic characteristic for the iron core, and the analysis was carried out with first order triangular elements as described in Section 5.2. The nonlinear iterations were performed by a simple chord method as well as by the Newton–Raphson technique employing the algorithm of Equation (5.7). From the resulting vector potential solution, the voltage induced in the winding is evaluated as follows.

The total instantaneous energy

$$W = \frac{1}{2} \iiint \mathbf{A} \cdot \mathbf{J} \, dxdydz. \tag{5.35}$$

The winding terminal emf is assumed to vary sinusoidally in time, so that

$$e = E \cos \omega t. \tag{5.36}$$

The total flux linkages λ for any coil are given by

$$\lambda = \int_0^t e \, dt. \tag{5.37}$$

Then the coil stored energy stated in terms of flux linkages and coil current i becomes

$$W = \frac{1}{2} \lambda i = \frac{1}{2} \lambda \iint \mathbf{J} \, dxdy. \tag{5.38}$$

Combining Equations (5.35)–(5.38) one obtains

$$\frac{1}{\omega} E \sin \omega t = \frac{\iiint \mathbf{A} \cdot \mathbf{J} \, dxdydz}{\iint \mathbf{J} \, dxdy}. \tag{5.39}$$

The results of the analysis were compared with search coil test data as shown in Figure 5.6. The flux distribution for this case is illustrated in Figure 5.7. Finally the wave form of the magnetizing current of the transformer was estimated and compared against test, as shown in Figure 5.8.

Andersen,[2] described a finite element solution of the transformer field problem in cylindrical polar coordinates, with θ directed currents in the windings.

Triangular finite elements of the first order were employed for the discretization of the field region. The approximate solution was defined such that the vector potential varies linearly in r, z. Thus

$$\mathbf{A} = \alpha_1 + \alpha_2 r + \alpha_3 z. \tag{5.40}$$

Equation (5.40) states that the radial component of flux density within each element is a constant. However, the axial component varies with the location,

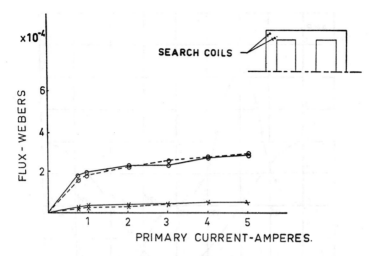

Figure 5.6 Variation of peak flux in corner region of transformer with primary current; curves correspond to peak flux densities of 0.8 Wb/m² and 2.15 Wb/m² respectively. —— test; ---- computed

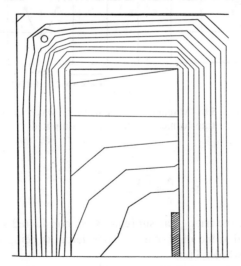

Figure 5.7 Predicted flux distribution in transformer with maximum flux density of 1.5 Wb/m² in core and 2.1 Wb/m² at corners

so that

$$\mathbf{B}_r = \frac{-\partial \mathbf{A}}{\partial z} = -\alpha_3 \tag{5.41}$$

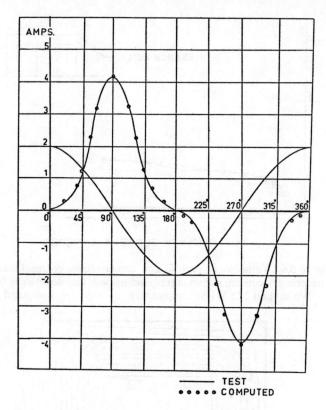

Figure   5.8 Magnetizing   current   waveform   of
transformer.——test; ..... computed

$$\mathbf{B}_z = \frac{\mathbf{A}}{r} + \frac{\partial \mathbf{A}}{\partial r} = \frac{\mathbf{A}}{r} + \alpha_2. \tag{5.42}$$

In this analysis, the triangular subregions are considered small enough such that the centroidal value of $\mathbf{B}_z$ can be considered as being representative for the triangle as a whole. Thus

$$\mathbf{B}_z = \frac{\mathbf{A}_c}{r_c} + \alpha_2. \tag{5.42a}$$

The total flux density is given by

$$B = \sqrt{(\mathbf{B}_r^2 + \mathbf{B}_z^2)} = \sqrt{\left\{\alpha_2^2 + \alpha_3^2 + \frac{2\alpha_2 \mathbf{A}_c}{r_c} + \left(\frac{\mathbf{A}_c}{r_c}\right)^2\right\}}. \tag{5.43}$$

The energy functional for the field problem was formulated as

$$W = \iint_R \left(\int_0^B vb\,db \right) r\,dr\,dz - \iint_R \mathbf{J} \cdot \mathbf{A} r\,dr\,dz. \tag{5.44}$$

Functional minimization was carried out in the usual manner for obtaining the vector potential solution.

The method was applied for evaluating reactances and forces in a transformer with windings of different lengths, an aluminium shield, unsymmetrical windings, and others. Flux plots obtained were compared to those obtained by Rabins.[33]

Silvester and Konrad[27] reformulated the functional of the partial differential equation of the field problem in terms of new variables defined as

$$\mathbf{u} = \mathbf{A}r^{-1/2} \tag{5.45}$$

$$\boldsymbol{\phi} = \mu_0 \mathbf{J} r^{-1/2}. \tag{5.46}$$

On substituting the variables $\mathbf{u}, \boldsymbol{\phi}$, and carrying out the differentiation, the energy functional becomes

$$\mathscr{F}(\mathbf{u}) = 2\pi \iint r^2 \left[\left(\frac{\partial \mathbf{u}}{\partial r} \right)^2 + \left(\frac{\partial \mathbf{u}}{\partial z} \right)^2 - 2\boldsymbol{\phi}\mathbf{u} \right] dr\,dz$$

$$+ \frac{10\pi}{4} \iint \mathbf{u}^2 dr\,dz + 6\pi \iint r\mathbf{u}\frac{\partial \mathbf{u}}{\partial r} dr\,dz. \tag{5.47}$$

$\boldsymbol{\mu}$ and $\boldsymbol{\phi}$ are expanded as polynomials of r, z in terms of the nodal values and area coordinates described by Zienkiewicz.[32] After lengthy algebra, the functional is reformulated as

$$\mathscr{F}(\mathbf{u}) = \frac{\pi}{2}(\mathbf{U}'\mathbf{S}\mathbf{U} - 2\mathbf{U}'\boldsymbol{T}\mathbf{V}) \tag{5.48}$$

where $\mathbf{U}$ and $\mathbf{V}$ are vectors of polynomial coefficients that describe the approximations of $\mathbf{u}$ and $\boldsymbol{\phi}$. $\mathbf{S}$ and $\boldsymbol{T}$ are square symmetric matrices.

Minimization of the energy functional results in a matrix equation of the form

$$[\mathbf{S}][\mathbf{U}] = [\boldsymbol{T}]. \tag{5.49}$$

The paper also describes the method of evaluating transformer reactance and forces on the windings. The stored energy in the leakage field is given by

$$W = \frac{1}{2} \int_R \mathbf{J} \cdot \mathbf{A}\,dR. \tag{5.50}$$

Reformulating Equation (5.50) in terms of the new variables $\mathbf{u}$ and $\boldsymbol{\phi}$

$$W = \pi \iint r^2 \mathbf{u} \boldsymbol{\phi} \, dr dz \tag{5.51}$$

so that in matrix quadratic form, W becomes

$$\mathbf{W} = \frac{\pi}{4} \mathbf{U}' \, \boldsymbol{T} \, \mathbf{V}. \tag{5.52}$$

Subsequently, the short-circuit reactance is calculated from the relation

$$X = \frac{2\omega \mathbf{W}}{I^2}. \tag{5.53}$$

The force on the windings is given by

$$\mathbf{F} = \int_R \mathbf{J} \times \mathbf{B} dR. \tag{5.54}$$

Substituting for $\mathbf{B}$ in terms of $\mathbf{A}$ and in turn by $\mathbf{u}$ and $\boldsymbol{\phi}$ and after considerable algebra, the z component of force is evaluated as

$$\mathbf{F}_z = \frac{2\pi}{\mu_0} \iint r^2 \phi \frac{\partial \mathbf{u}}{\partial z} dr dz. \tag{5.55}$$

Di Monaco, Giuseppetti and Tontini[14] described a similar analysis and its application for evaluating transformer magnetic fields. It is reported that triangular, quadrilateral, and isoparametric elements were used in their analysis. Extensive comparison between numerical results and tests were presented.

5.5 ANALYSIS OF THE EDDY-CURRENT PROBLEM

Silvester and Haslam[25] first presented a finite element analysis of eddy-current fields in magnetotelluric problems. Chari[11] showed a similar formulation for the eddy-current problem in magnetic structures, described hereunder. The governing partial differential equation and the functional formulation in a two-dimensional cartesian system were expressed as

$$\frac{\partial^2 \mathbf{A}}{\partial x^2} + \frac{\partial^2 \mathbf{A}}{\partial y^2} = \frac{j\omega\mu\mu_0 \mathbf{A}}{\rho} - \mathbf{J}_s \tag{5.56}$$

where $\mathbf{A}$ is assumed to be a periodic function of time, so that

$$\mathscr{F} = \iint \frac{1}{2} \left\{ \left(\frac{\partial \mathbf{A}}{\partial x} \right)^2 + \left(\frac{\partial \mathbf{A}}{\partial y} \right)^2 \right\} dxdy + \frac{j\omega\mu\mu_0}{2\rho} \iint \mathbf{A}^2 dxdy - \iint \mathbf{J}_s \cdot \mathbf{A} dxdy. \tag{5.57}$$

The minimization of the above functional for first order triangular elements, leads to a complex matrix equation of the form

$$[S][A]+[C][A]=[T] \tag{5.58}$$

where $[T]$ is a forcing function, $[S]$ is a real symmetric matrix identical to that of Equation (5.6), and $[C]$ is a complex symmetric matrix of the form

$$[C]=\frac{j\omega\mu\mu_0\Delta}{12\rho}\begin{bmatrix}2 & 1 & 1\\ 1 & 2 & 1\\ 1 & 1 & 2\end{bmatrix}.$$

The analysis was applied to semi-infinite and finite conducting plates and the results were compared with tests.

A useful analysis of the axisymmetric electromagnetic induction problem was presented in Reference 15. On the assumption of a θ-directed vector potential $A_\theta(r, z)$, the following formulation was obtained for the field problem:

$$\frac{1}{\mu}\left(\frac{\partial^2 A}{\partial r^2}+\frac{1}{r}\frac{\partial A}{\partial r}+\frac{\partial^2 A}{\partial z^2}-\frac{A}{r^2}\right)+(\omega^2\epsilon-j\omega\sigma)A$$
$$+\frac{\partial(1/\mu)}{\partial r}\left(\frac{1}{r}\frac{\partial(rA)}{\partial r}\right)+\frac{\partial(1/\mu)}{\partial z}\frac{\partial A}{\partial z}+J_0=0. \tag{5.59}$$

The variational expression for the above was shown to be

$$\mathcal{F}=\iiint_D\frac{1}{\mu}(\nabla A.\nabla\delta A)dD-\iiint_D\left\{\left(\omega^2\epsilon-\frac{1}{\mu r^2}-j\omega\sigma\right)\right\}A\delta AdD$$
$$-\iiint_D\left\{\left[\frac{\partial(1/\mu)}{\partial r}\left(\frac{1}{r}\frac{\partial(rA)}{\partial r}\right)+\frac{\partial(1/\mu)}{\partial z}\frac{\partial A}{\partial z}\right]\delta A\right\}dD \tag{5.60}$$
$$-\iiint_D J_0\delta AdD-\iint_s\frac{1}{\mu}\frac{\partial A}{\partial n}\delta Ads=0.$$

The functional was minimized in the usual manner and extended to nonlinear media. An application of the method to a coil surrounding a conducting rod was presented, and the magnitude and phase of the vector potential were plotted.

Foggia, Sabonnadière, and Silvester[16] presented a finite element solution of saturated travelling magnetic field problems in a linear induction motor. The partial differential equation for the diffusion problem was modified to include a translation term, so that

$$\frac{\partial}{\partial x}\left(\frac{1}{\mu}\frac{\partial A}{\partial x}\right)+\frac{\partial}{\partial y}\left(\frac{1}{\mu}\frac{\partial A}{\partial y}\right)=\frac{1}{\rho}\left(\frac{\partial A}{\partial t}+V\frac{\partial A}{\partial x}\right) \tag{5.61}$$

where **V** is the translation velocity.

Instead of the conventional procedure of formulating an energy functional and minimizing the same for obtaining the field solution, the authors presented a (weak) Galerkin formulation with the result:

$$\int_\Omega \left(\frac{\partial \mathbf{A}}{\partial t} + \mathbf{V}\frac{\partial \mathbf{A}}{\partial x}\right)\phi(x,\ y)\mathrm{d}x\mathrm{d}y + \int_\Omega \rho v\left(\frac{\partial \mathbf{A}}{\partial x}\cdot\frac{\partial \mathbf{V}}{\partial x} + \frac{\partial \mathbf{A}}{\partial y}\cdot\frac{\partial \mathbf{V}}{\partial y}\right)\mathrm{d}x\mathrm{d}y = 0 \quad (5.62)$$

where ϕ is required to vanish wherever essential boundary conditions apply to **A** on the boundary Γ. The vector potential at any interior point is given by

$$\mathbf{A}_n = \sum_{j=1}^{3} L_j(x,\ y)\mathbf{A}_j(t). \quad (5.63)$$

In Equation (5.63), the index j ranges over the vertices of the triangle. The functions L_j are identical to the interpolation functions α_j described by Silvester.[22]

Substituting for **A** from Equation (5.63) in Equation (5.62), and substituting for

$$\phi(x,\ y) = \sum_{j=1}^{3} L_j(x,\ y) \quad (5.64)$$

a matrix equation is obtained of the form

$$[\boldsymbol{M}]\frac{\mathrm{d}}{\mathrm{d}t}[\mathbf{A}] + [\boldsymbol{N}][\mathbf{A}] = \mathbf{0}. \quad (5.65)$$

Equation (5.65) was solved step-by-step with small time increments, using the recursive relation of the nonlinear equation

$$\{\mathbf{A}_{n+1}^{K+1}\} = (\{\boldsymbol{M}\} + \Delta t\{\boldsymbol{N}_{n+1}^{K}\})^{-1}\{\boldsymbol{M}\}(\mathbf{A}_n). \quad (5.66)$$

The solution of Equation (5.66) at each step was performed by gaussian elimination. The reluctivities were under-relaxed to ensure convergence. The analysis was applied to an experimental linear induction motor and the air-gap flux distribution, levitation force, and drag force were calculated and compared with test results.

Carpenter[7] described a finite element network model and its application to eddy-current problems. It is stated that since the dissipation term affects the variational formulation, it is more accurate to form a network model from the linked current and flux patterns.

Csendes and Chari[13] describe a general variational formulation for the diffusion problem including the rotational term. This generalization of the eddy-current problem is well suited for analysing the asynchronous performance of rotating electrical machines. In this method, the partial differential equation and its functional formulation are given by

$$v\nabla^2 A = \frac{1}{\rho} \frac{\partial A}{\partial t} + V \times B - J_s \qquad (5.67)$$

$$\mathscr{F} = \int_R |\nabla A|^2 dR + \frac{\mu\omega}{\rho} \int_R \left(y \frac{\partial A}{\partial x} - x \frac{\partial A}{\partial y} \right) dR - j \frac{\mu\omega}{\rho} \int_R A^2 dR - 2 \int_R A \cdot J dR. \qquad (5.68)$$

This technique was applied to a simplified generator cross-section discretized by first order triangular elements, and a magnitude plot of the flux was presented.

Hsieh and Silvester[24] described a method of including the exterior boundary by means of a free space Green's function. Sabonnadière, Coulomb, Silvester, and Konrad[19] solved a skin effect problem by separating the time and space dependence from the variational formulation. The equation dependent on time is then solved by the use of periodic functions. The space-dependent equation is solved as an eigenvalue problem. The two solutions are then coupled to yield the required solution to the field problem.

5.6 SCALAR POTENTIAL SOLUTIONS BY 2-D AND 3-D FINITE ELEMENTS

Andersen[3] described a two-dimensional scalar potential solution for the electrostatic field evaluation in electrical machines. The functional formulation and its minimization are obtained in a similar manner to the magnetic field analysis, employing an iterative equation solution procedure. An automatic grid generator program is described for rectilinear and curved boundaries. A good correlation between the results of the finite element analysis and conventional analysis was reported.

Armor and Chari[4] described a three-dimensional scalar potential solution of the steady state heat conduction problem in the stator core of a turbine generator. Perspective isothermal plots were presented and correlation with test results were illustrated. In this analysis, triangular, rectangular, and trapezoidal prism elements were employed.

5.7 CONCLUSIONS

A description of the finite element method and its application to diverse field applications in electrical machines are presented. The variational formulations for Laplace, Poisson, and diffusion equations, which govern the field problems, are furnished. The specification of the approximating function and functional minimization are illustrated by employing first order triangular finite elements. Solutions obtained for two- and three-dimensional electromagnetic, diffusion, static, and thermal field problems in cartesian and cylindrical polar coordinates are surveyed.

The aforesaid method, based on a variational formulation of the partial differential equation in terms of an energy functional and the minimization of the latter to obtain the required solution, has been used in the last few years for a variety of field problems.

By virtue of implicit natural boundary conditions and efficient equation solving algorithms, the finite element technique is proving to be highly accurate, economical, and a useful design tool for machine analysis.

REFERENCES

1. Andersen, O. W. (1972). 'Iterative solution of finite element equations in magnetic field problems', *IEEE Power Engineering Society, Paper T* 72 411–7.
2. Andersen, O. W. (1973). 'Transformer Leakage Flux Program based on the Finite Element Method', *IEEE Trans.*, **PAS-92**, No. 2.
3. Andersen, O. W. (1973). 'Laplacian Electrostatic Field Calculations by Finite Elements with Automatic Grid Generation', *IEEE Trans.*, **PAS-92**, No. 5.
4. Armor, A. F. and Chari, M. V. K. (1976). 'Heat Flow in the stator core of large Turbine-Generators, by the method of Three-dimensional Finite Elements', *IEEE Winter Meeting, New York, Paper No. F* 76 027–3.
5. Brandl, P., Reichert, K., and Vogt, W. (1975). 'Simulation of Turbogenerators on Steady State Load', *Brown Boveri Review*, **9**.
6. Brauer, J. R. (1975). 'Saturated Magnetic Energy Functional for Finite Element Analysis of Electric Machines', *Paper C75-151-6 presented at IEEE Power Engineering Society Winter Meeting, New York City, January 28, 1975*.
7. Carpenter, C. J. (1975) 'Finite-element network models and their application to eddy-current problems', *Proc. IEEE*, **122**, No. 4.
8. Chari, M. V. K. (1970). 'Finite Element Analysis of Nonlinear Magnetic Fields in Electric Machines', *Ph.D. Dissertation*, McGill University, Montreal, Canada.
9. Chari, M. V. K. and Silvester, P. (1971). 'Analysis of Turbo-alternator magnetic fields by Finite Elements', *IEEE Trans.*, **PAS-90**, 454–464.
10. Chari, M. V. K., Sharma, D. K., and Kudlacik, H. W. (1976). 'No load magnetic field analysis in the end region of a turbine generator by the method of finite elements', *IEEE Winter Meeting, New York, Paper No. A* 76 230–3.
11. Chari, M. V. K. (1973). 'Finite-Element Solution of the Eddy-current Problem in Magnetic Structures', *IEEE Trans.*, **PAS-92**, No. 1.
12. Colonias, J. S. (1974). *Particle Accelerator Design Computer Programs*, Academic Press, New York.
13. Csendes, Z. J. and Chari, M. V. K. (1976). 'General Finite Element Analysis of Rotating Electric Machines', *International Conference on numerical methods in Electric and magnetic field problems, St. Margherita, Italy*.
14. Di Monaco, A., Giuseppetti, G., and Tontini, G. (1975). 'Studio Di Campi Elettrici E Magnetici Stanzionari Con Il Metodo Degli Elementi Finiti—Applicazione Ai Transformatori', *L'Elettrotecnica*, **LXII**, No. 7, 585–598.
15. Donea, J., Giuliani, S., and Philippe, A. (1974). 'Finite Elements in the solution of Electromagnetic induction problems', *International Journal for numerical methods in Engineering*, **8**, 359–367.
16. Foggia, A., Sabonnadière, J. C., and Silvester, P. (1975). 'Finite Element Solution of Saturated Travelling Magnetic Field problems', *IEEE Trans.*, **PAS-94**, No. 3.
17. Glowinski, R. and Marrocco, A. (1974). 'Analyse Numerique du champ magnetique d'un Alternateur par elements finis et surrelaxation ponctuelle non lineaire', *Computer methods in Applied Mechanics and Engineering*, **3**, 55–85.

18. Howe, D. and Hammond, P. (1974). 'The distribution of axial flux on the stator surface of the ends of turbogenerators', *Proc. IEE*, **121**, No. 9, 980–990.
19. Konrad, A., Coulomb, J. L., Sabonnadière, J. C., and Silvester, P. P. (1976). 'Finite Element Analysis of Steady-State skin effect in a slot-embedded conductor', *IEEE Winter Meeting, New York, Paper No. A 76 189–1.*
20. Kreisinger, V. (1974). 'Iterative methods for the solution of nonlinear magnetic fields', *ACTA TECHNICA CSAV*, **3**.
21. Okuda, H., Kawamura, T., and Nishi, M. (1976). 'Finite-Element Solution of Magnetic Field and Eddy Current Problems in the End Zone of Turbine Generators', *IEEE Winter Meeting, New York, Paper No. A 76 141–2.*
22. Silvester, P. (1969). 'High-Order Polynomial Triangular Finite Elements for Potential Problems', *Int. Journal of Engineering Science*, **7**, 849–861.
23. Silvester, P. and Chari, M. V. K. (1970). 'Finite Element Solution of Saturable Magnetic Field Problems, *IEEE Transactions on Power Apparatus and Systems*, **PAS-89**, No. 7, 1642–1651.
24. Silvester, P. and Hsieh, M. S. (1971). 'Finite-Element Solution of 2-Dimensional Exterior Field Problems', *Proc. IEE*, **118**, No. 12, 1743–1747.
25. Silvester, P. and Haslam, C. R. S. (1972). 'Magnetotelluric Modelling by the Finite Element Method', *Geophysical Prospecting*, **20**, 872–891.
26. Silvester, P., Cabayan, H. S., and Browne, B. T. (1973). 'Efficient Techniques for Finite Element Analysis of Electric Machines', *IEEE Trans.*, **PAS-92**, No. 4.
27. Silvester, P. and Konrad, A. (1973). 'Analysis of Transformer Leakage Phenomena by High-Order Finite Elements', *IEEE Trans.*, **PAS-92**, No. 6.
28. Silvester, P. and Rafinejad, P. (1974). 'Curvilinear Finite Elements for Two-Dimensional Saturable Magnetic Fields', *IEEE Trans.*, **PAS-93**, No. 6.
29. Unterweger, P. (1973). 'Computation of Magnetic Fields in Electrical Apparatus', *IEEE Summer Meeting, Paper No. T 73 352–2.*
30. Wexler, A. (1973). 'Finite-Element Analysis of Inhomogeneous Anisotropic Reluctance Machine Rotor', *IEEE Trans.*, **PAS-92**, No. 1, 145–149.
31. Winslow, A. M. (1965). 'Magnetic Field Calculations in an Irregular Triangular Mesh', *Lawrence Radiation Laboratory*, Livermore, California, UCRL-7784-T, Rev. 1.
32. Zienkiewicz, O. C. (1971). *The Finite Element Method in Engineering Science, McGraw Hill, London.*
33. Rabins, L. (1955). 'Transformer Reactance Calculations with Digital Computers', A.I.E.E. Winter meeting, New York 1955.

Finite Elements in Electrical and Magnetic Field Problems
Edited by M. V. K. Chari and P. P. Silvester
© 1980, John Wiley & Sons Ltd.

Chapter 6

Numerical Analysis of Eddy-Current Problems

J. C. Sabonnadière

6.1 INTRODUCTION

Eddy currents have for long been considered as harmful, to be kept to the minimum possible because they produce losses, thus increasing the thermal requirements and decreasing efficiency. They have been reduced by subdividing conductors, for example, by lamination of magnetic ciruits in alternating current electric machines, or Roebel bars and hollow conductors in high-power a.c. generators.

When electric machines are fed by systems using thyristors and other power electronics devices, the high-frequency components of current and voltage increase, thus iron losses (especially eddy-current losses) are substantially augmented. It then becomes important to take account of these losses in the design of the machines. Another source of annoyance created by eddy currents is the electrodynamic forces between windings or different current-carrying parts of apparatus (e.g. bus bars).

The prevention of this kind of undesirable effect in modern apparatus requires a good knowledge of the eddy-current distribution and of related phenomena, such as skin effect or reaction forces.

Furthermore, new technologies have recently arisen which make use of eddy currents for production of forces or heat. We could mention for example the linear induction motor, in which eddy currents are produced in a reaction rail to create a propulsive force. Another example arises in electrodynamic suspension systems, in which eddy currents provide the lift force but unfortunately create a drag force also. In induction heating, the principle is to induce Joule losses in the workpiece; the losses are generated by eddy currents.

The design of propulsion and suspension systems, or of induction heating, requires a much greater accuracy in eddy-current computations than previously, because in these devices eddy currents are the basis phenomenon on which machine operation depends. Generally a two-dimensional model may furnish a sufficiently good approach for engineering purposes, but sometimes three-dimensional analysis becomes necessary.

An analysis of the methods used for the calculation of eddy currents must be related closely to the structures in which these currents are generated. Instead of a rough three-dimensional model, engineers often prefer sophisticated two-dimensional models which, when correctly stated, give an accuracy consistent with the design objective.

The presentation below consists of two main parts: the structures, and the problems to be solved including indications on the methods of solution.

6.2   THE STRUCTURES IN WHICH EDDY CURRENTS ARE DEVELOPED

Industrial structures in which we find eddy currents are essentially of two kinds: long structures, in which the electric field and current density possess only one component; and broad structures, in which currents can flow in two or three directions.

6.2.1   Long structures

In such structures, currents follow paths determined by the nature of the conductor structure. The voltage applied at the conductor terminals may be generated either by an electric field applied at these terminals, or by a time-varying magnetic flux linking the loop formed by the conductors. This kind of problem is encountered in various areas.

Conductors, bus bars, overhead lines

These devices are essentially components of electric transmission networks (overhead lines) or of distribution networks (bus bars, underground cables, etc.).

In this kind of problem we know the applied voltage at line, cable, or bus

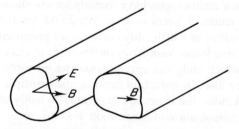

Figure   6.1   Structures   with   one-dimensional electric field

bar terminals, and we seek the current density distribution within the conductor in order to compute the electrodynamic forces, total losses, or (more generally) electromagnetic parameters like self or mutual inductances, taking the skin effect into account, especially in transient (e.g. short-circuit) operation.

Squirrel cage or damper windings for a.c. machines

Here we have short-circuited multiconductor systems in which the induced currents are generated by application of a time-varying flux through the system. In this case, the current distribution is needed for impedance calculation and loss evaluation at various frequencies (asynchronous operation for an alternator, variable-speed drive in the case of an induction motor) for which the skin effect is not negligible.

Distributions which have axial symmetry can be analysed by introducing a magnetic vector potential which has the same direction as the current density. We shall investigate this kind of problem in more detail in the next section.

6.2.2 Broad structures

This type of structure possesses, from the industrial point of view, the great advantage of simplicity of construction. Conversely, it leads to complexity in its modelling and in calculation.

Reaction rail of induction machines

Generally two-dimensional because its small thickness makes skin effect negligible, this problem is of great importance for linear induction motor engineering. As a matter of fact, the distribution of the current density vector in the plane of the rail is critical to the operation of the motor, because the propulsive force is the resultant of its cross-product with the air-gap field.

Electric heating by electromagnetic induction

This application of eddy currents is one in which it is necessary to deal accurately with the generation of a given eddy-current distribution. Indeed, if the eddy-current distribution is highly non-uniform, the resulting high temperature gradients may cause the workpiece to crack. This problem is a very difficult one to solve because it requires a three-dimensional analysis of the Helmholtz equation, occasionally including nonlinear behaviour of the material.

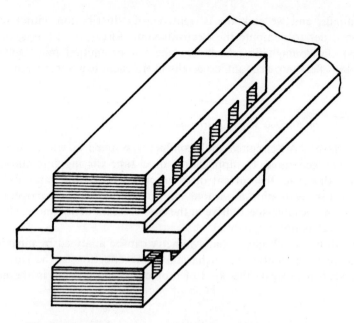

Figure 6.2   Broad structures: the linear induction machine

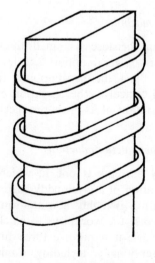

Figure 6.3   A three-dimensional magnetic heating
problem

Magnetic levitation systems

Here we have, in general, an infinitely long rod of high-conductivity material, in which induced currents are created by the movement of a magnetic field source in the vicinity of the rod. The field, generally constant in time, varies with space position and the translation of the source produces, by speed effect, a variation of the flux linking the rod. This problem is also three-dimensional, but may be reduced to a two-dimensional problem by the use of Fourier series or Fourier integrals, combined with a termwise two-dimensional field analysis.

As was briefly stated above, problems of eddy-current analysis are different in the two different classes of structure; furthermore, in either case they differ for currents, i.e. flux sources, varying in time (a.c. current), or by way of the motion of a space-variable time-invariant field source. For any given application (electric machines, induction heating) the presence of saturable iron sheets or castings introduces saturation phenomena into the analysis, and makes the problem nonlinear.

For each case we can apply general mathematical methods, but commonsense requires practical computation algorithms to be fitted to each kind of problem. For this reason, we shall divide the general problem of eddy current analysis into several subproblems, each derived from the general problem by additional physical assumptions suited to it.

6.3 THE PROBLEM

In a first step we shall consider the general problem of the distribution of eddy currents in a slab of conducting saturable material. Let $\mathbf{B}_{ex}$ (x, y, z, t) be the time-varying and space-varying excitation magnetic flux density in a domain of the three-dimensional euclidean space in which there lies a slab of conducting material. Let the following notation be introduced for vectors, each a function of x, y, z, and t.

$\mathbf{B}_r$ reaction magnetic flux density
$\mathbf{B}$ total magnetic flux density
$\mathbf{H}_r$ reaction magnetic field
$\mathbf{H}$ total magnetic field
$\mathbf{E}$ induced electric field
$\mathbf{E}_{app}$ externally applied electric field
$\mathbf{J}$ current density inside the conducting material
σ conductivity
μ magnetic permeability
v magnetic reluctivity
$\mathbf{v}$ relative velocity between the magnetic field source and the slab.

The Maxwell equations of electrodynamics may be written, neglecting displacement currents, as

$$\mathbf{V} \times \mathbf{E} + \frac{d}{dt}(\mathbf{B}_r + \mathbf{B}_{ex}) = 0 \tag{6.1}$$

$$\mathbf{V} \times \mathbf{H} = \mathbf{J} \tag{6.2}$$

$$\mathbf{J} = \sigma(\mathbf{E} + \mathbf{E}_{app}) \tag{6.3}$$

$$\mathbf{B} = \mu(|\mathbf{H}|)\mathbf{H} \tag{6.4}$$

$$\mathbf{V} \cdot \mathbf{B} = 0 \tag{6.5}$$

$$\mathbf{V} \cdot \mathbf{J} = 0 \tag{6.6}$$

$$\mathbf{B} = \mathbf{B}_{ex} + \mathbf{B}_r. \tag{6.7}$$

In these equations we have made an explicit distinction between the excitation flux density and the reaction field flux density. In a similar way, we consider the electric field to be the sum of two vectors, the induced electric field $\mathbf{E}$ and the externally applied electric field $\mathbf{E}_{app}$. This distinction is often useful in further steps in the solution of these equations.

Some boundary conditions must be added to these partial differential equations to provide a mathematically well-posed problem descriptive of the physical phenomenon under analysis. From experience, we can state that the most frequent boundary conditions occurring in these kind of problems are Neumann or Dirichlet conditions. Occasionally, interface continuity conditions arise at interfaces between different media ($\mu_1 \mathbf{H}_{n,1} = \mu_2 \mathbf{H}_{n,2}$).

In its full generality, this problem has to be solved in four dimensions (three space dimensions and one for time variation), and it is strongly nonlinear. As a matter of fact, if the material is saturable, its permeability depends on the magnetic field strength and thus on the solution itself.

To try to obtain a general solution for so complicated a problem appears quite unrealistic, and one is well-advised to seek simplifications which reduce both the generality and the difficulty of the problem so as to obtain less rigorous but more directly computable solutions to industrial problems. Thus it is usually, if not indeed always, the rule in engineering to classify problems in categories according to their mathematical formulations and their physical meaning.

6.3.1   Two-dimensional linear problems

In this class of problem, saturation is neglected and assumptions are made which lead to the search for solutions in a two-dimensional space.

Long structures

Here the electric field is assumed to be z-directed (Figure 6.4), while the magnetic field lies in the cross-sectional plane of the conductor. Introduction of the vector potential $\mathbf{A}$, which has a single component along the z-axis, allows reduction of the problem to a scalar one. In accordance with Coulomb[1,2] we introduce the following notation:

$v_r(x, y) = \mu_0/\mu(x, y)$, the relative reluctivity.

$\sigma_r(x, y) = \sigma(x, y)/\sigma_0$, relative conductivity, with σ_0 a constant reference conductivity.

$\mathbf{A}$ is the z-component of the vector potential.

$\mathbf{E}_0(t)$ is the externally applied electric field.

With the assumptions and notation above, Maxwell's equations lead to

$$\frac{\partial}{\partial x}\left[v_r(x, y)\frac{\partial \mathbf{A}(x, y, t)}{\partial x}\right]+\frac{\partial}{\partial y}\left[v_r(x, y)\frac{\partial \mathbf{A}}{\partial y}(x, y, t)-\mu_0\sigma_0\sigma_r(x, y)\frac{\partial \mathbf{A}}{\partial t}(x, y, t)\right]=$$
$$\mu_0\sigma_0\sigma_r(x, y)\mathbf{E}_0(t).$$

$$(6.8)$$

The current density satisfies the scalar equation

$$\mathbf{J}_z(x, y, t) = -\frac{1}{\mu_0}\left[\frac{\partial}{\partial x}\left(v_r(x, y)\frac{\partial \mathbf{A}}{\partial x}\right)+\frac{\partial}{\partial y}\left(v_r(x, y)\frac{\partial \mathbf{A}}{\partial y}\right)\right]. \qquad (6.9)$$

Boundary conditions applicable on the boundaries Γ of the domain are

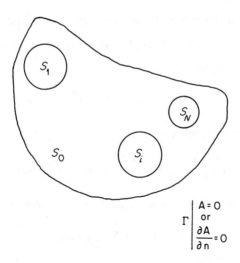

Figure 6.4 Domain for analysis of long structures

homogeneous Neumann $(\partial \mathbf{A}/\partial n = 0)$ or homogeneous Dirichlet $(\mathbf{A} = 0)$ conditions.

The best way of solving this problem is to consider it as an eigenvalue problem. To this end, rewrite Equation (6.8) with the right-hand side set to zero:

$$\frac{\partial}{\partial x}\left(v_r \frac{\partial \mathbf{A}}{\partial x}\right) + \frac{\partial}{\partial y}\left(v_r \frac{\partial \mathbf{A}}{\partial y}\right) - \mu_0 \sigma_0 \sigma_{0r} \frac{\partial \mathbf{A}}{\partial t} = 0. \tag{6.10}$$

Proceeding by separation of variables, we seek a solution of the type $\mathbf{A}(x, y, t) = \mathbf{\alpha}(x, y,)\tau(t)$. Equation (6.10) then becomes

$$\tau\left[\frac{\partial}{\partial x}\left(v_r \frac{\partial \mathbf{\alpha}}{\partial x}\right) + \frac{\partial}{\partial y}\left(v_r \frac{\partial \mathbf{\alpha}}{\partial y}\right)\right] = \mu_0 \sigma_0 \sigma_r \mathbf{\alpha}\frac{d\tau}{dt}. \tag{6.11}$$

After separation and introduction of the separation constant λ,

$$\mu_0 \sigma_0 \frac{d\tau}{dt} = -\lambda\tau \tag{6.12}$$

$$\frac{\partial}{\partial x}\left(v_r \frac{\partial \mathbf{\alpha}}{\partial x}\right) + \frac{\partial}{\partial y}\left(v_r \frac{\partial \mathbf{\alpha}}{\partial y}\right) = -\lambda \sigma_r \mathbf{\alpha}. \tag{6.13}$$

The latter equation has an infinite number of eigenvalues λ_k and associated eigenfunctions $\mathbf{\alpha}_k(x, y)$, which are orthogonal over the domain Ω,

$$\int_\Omega \sigma_r(x, y)\mathbf{\alpha}_k(x, y)\,\mathbf{\alpha}_j(x, y)\mathrm{d}x\mathrm{d}y = 0 \qquad \text{if } k \neq j, \tag{6.14}$$

Now assume $\mathbf{A}$ to be expressible as the sum of two functions:

$$\mathbf{A}(x, y, t) = \mu_0 \sigma_0 \mathbf{E}_0(t)\mathbf{\alpha}_0(x, y) + \mathbf{A}_c(x, y, t) \tag{6.15}$$

where $\mathbf{\alpha}_0$ is the solution of

$$\frac{\partial}{\partial x}\left(v_r \frac{\partial \mathbf{\alpha}_0}{\partial x}\right) + \frac{\partial}{\partial y}\left(v_r \frac{\partial \mathbf{\alpha}_0}{\partial y}\right) = -\sigma_r. \tag{6.16}$$

Then the correction term $\mathbf{A}_c$ satisfies

$$\frac{\partial}{\partial x}\left(v_r \frac{\partial \mathbf{A}_c}{\partial x}\right) + \frac{\partial}{\partial y}\left(v_r \frac{\partial \mathbf{A}_c}{\partial y}\right) - \mu_0 \sigma_0 \sigma_r \frac{\partial \mathbf{A}_c}{\partial t} = \mu_0^2 \sigma_0^2 \sigma_r \frac{d\mathbf{E}_0}{dt}\,\mathbf{\alpha}_0. \tag{6.17}$$

In terms of the eigenfunctions $\mathbf{\alpha}_k(x, y)$, $\mathbf{A}_c$ may be written

$$\mathbf{A}_c(x, y, t) = \sum_{j=1}^{\infty} a_j(t)\,\mathbf{\alpha}_j(x, y) \tag{6.18}$$

where the unknown functions a_j are solutions of the differential equation resulting from the separation Equations (6.12, 6.13)

$$-\sum_{j=1}^{\infty}\left[\frac{\lambda_j}{\mu_0\sigma_0}\,a_j+\frac{da_j}{dt}\right]\mathbf{a}_r\,\mathbf{a}_j=\mu_0\sigma_0\sigma_r\mathbf{a}_0\,\frac{dE_0}{dt}. \tag{6.19}$$

Multiplication of both sides of this equation by the eigenfunction $\mathbf{a}_k(x,\,y)$ and integration over Ω yields for a_k

$$-\left|\frac{\lambda_k}{\mu_0\sigma_0}\,a_k+\frac{da_k}{dt}\right|\int_{\Omega}\sigma_r a_k^2\,d\Omega=\mu_0\sigma_0\,\frac{dE_0}{dt}\int_{\Omega}\sigma_r\mathbf{a}_0\mathbf{a}_r\,d\Omega. \tag{6.20}$$

The solution of this equation gives the vector potential $\mathbf{A}_c$ and thus, by (6.16) and (6.18), the vector potential $\mathbf{A}$. However, an alternative procedure may be obtained by expanding $\mathbf{a}_0$ in the eigenfunctions $\mathbf{a}_k$,

$$\mathbf{a}_0(x,\,y)=\sum_{j=1}^{n}u_j\mathbf{a}_j(x,\,y). \tag{6.21}$$

Since $\mathbf{a}_0$ satisfies Equation (6.16),

$$\sum_{j=1}^{\infty}u_j(-\lambda_j\sigma_r\mathbf{a}_j)=-\sigma_r. \tag{6.22}$$

Multiplication by $\mathbf{a}_k$ and integration over Ω allows computation of the coefficients u_k

$$u_k=\frac{\int_{\Omega}\sigma_r\mathbf{a}_k d\Omega}{\lambda_k\int_{\Omega}\sigma_r a_k^2 d\Omega}. \tag{6.23}$$

Thus for $k=1,\,2,\ldots,\,n$, the functions a_k are solutions of the differential equations

$$\frac{\lambda_k}{\mu_0\sigma_0}\,a_k(t)+\frac{da_k}{dt}=-\frac{\mu_0\sigma_0}{\lambda_k}\,\frac{dE_0}{dt}\,\frac{\int_{\Omega}\sigma_k\mathbf{a}_k d\Omega}{\int_{\Omega}\sigma_k a_k^2 d\Omega}. \tag{6.24}$$

The current density, given by Equation (6.9) can be written in terms of the eigenfunctions $\mathbf{a}_k$ and the time functions a_k:

$$\mathbf{J}(x,\,y,\,t)=\sigma_0\sigma_r(x,\,y)\mathbf{E}_0(t)+\sum_{k=1}^{\infty}\lambda_k\sigma_r(x,\,y)\mathbf{a}_k(x,\,y)a_k(t). \tag{6.25}$$

In expression (6.25) we note that $\mathbf{J}$ may be regarded as decomposable into a term $\mathbf{J}_0=\sigma_0\sigma_r\mathbf{E}_0(t)$ obtained when we neglect the skin effect, and the infinite summation of terms $\mathbf{J}_k=(\lambda_k/\mu_0)\sigma_r\mathbf{a}_r(x,\,y)a_k(t)$ due to skin effect.

Numerical methods are employed for the eigenvalue and eigenfunction calculations. Use of the finite element method implies a variational formulation of the eigenvalue problem. The functional corresponding to the homogeneous Helmholtz equation is

$$F(\alpha) = \frac{1}{2} \int_{\Omega} \left[v_r \left(\frac{\partial \alpha}{\partial x} \right)^2 + v_r \left(\frac{\partial \alpha}{\partial y} \right)^2 - \lambda \sigma_r \alpha^2 \right] d\alpha. \qquad (6.26)$$

Decomposing Ω into finite elements Ω_i, the functions α are sought in terms of a polynomial approximation of each element:

$$\alpha_i(x, y) = \sum_j b_j(x, y) \mathbf{A}_j \qquad (6.27)$$

where $b_j(x, y)$ are polynomials and $\mathbf{A}_j$ are unknown coefficients which will be determined by the values of α_i at the nodes of the element. The extremum requirement on the functional $F(\alpha)$, combined with the topological relations between elements, transforms the differential equation eigenvalue problem into an equivalent algebraic eigenvalue problem:

$$| \; S - \lambda \; T | \; |\mathbf{A}| = 0 \qquad (6.28)$$

in which the vector $\mathbf{A}$ is the column of the coefficients $\mathbf{A}_j$, while the matrices S and T are built up from the elementary coefficients

$$S^i_{kj} = \int_{\Omega_i} v_r \left(\frac{\partial b_k}{\partial x} \frac{\partial b_j}{\partial x} + \frac{\partial b_k}{\partial y} \frac{\cdot b_j}{\partial y} \right) dx dy \qquad (6.29)$$

$$T^i_{kj} = \int_{\Omega_i} \sigma_r b_k b_j dx dy \qquad (6.30)$$

by the assembly of elements Ω_i.

The eigenvalues of the original partial differential equation problem are approximated by the eigenvalues λ_k obtained by solving this discrete eigenvalue problem. In the same way, the eigenvectors $\mathbf{A}_k$ of the system (6.28) allow computation of the eigenfunctions α_k of the continuum problem by the polynomial approximation (6.27) on each element.

This method has been tested on special problems where analytic calculations were possible, and the results have been compared with theoretical results. The tests show in each case good agreement between numerical and theoretical methods. Furthermore, the numerical method was tested on a T-shaped conductor, and the results fitted experimental results very well for a large range of frequencies.

This method has been developed elsewhere[3] and is believed to be applicable to a large number of electrical engineering problems.

Broad structures

We will here study the distribution of eddy currents in a rail of infinite length and finite width. If we assume the current distribution to be two-dimensional, the flux density must be assumed to lie in a plane orthogonal to the rail. The best method for computing the field and current is then to compute accurately the primary field and afterwards to make a spatial Fourier analysis of the current density components in the plane of the rail. This method is easy to apply because the shape of the rail is generally very simple (nearly always a rectangle or an infinitesimally thin strip), so the solution of the Poisson or Helmholtz equation is feasible by eigenvalue analysis or by variable separation.[4,5]

6.3.2 Quasi-three-dimensional linear problems

In linear motor problems it is not always possible to use Fourier decomposition, because accuracy requirements may impose a three-dimensional

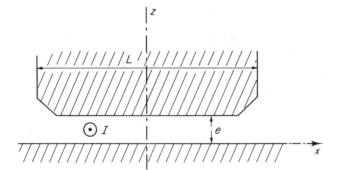

Figure 6.5 Linear motor, separated by air-gap from infinite rail

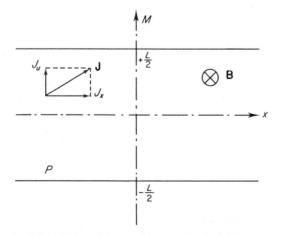

Figure 6.6 Linear motor rail, top view

analysis which takes into account the exact shape of windings. In this case it is possible to develop an iterative method, using more sophisticated field analysis techniques such as Green's functions or conformal mapping. It will be assumed here that the vector $\mathbf{J}$ of the current density lies in the plane of the reaction rail, while the flux density is directed along the z axis.

The algorithm used to compute eddy currents in the rail and the reaction field in the air gap is based upon the Maxwell equations (6.31) to (6.34):

$$\mathbf{V} \times \mathbf{J} = -\left(\frac{\partial \mathbf{B}}{\partial t} - v\frac{\partial \mathbf{B}}{\partial x}\right)\sigma \tag{6.31}$$

$$\text{div } \mathbf{B} = 0 \tag{6.32}$$

$$\mathbf{V} \times \mathbf{B} = \mu_0 \mathbf{J} \tag{6.33}$$

$$\text{div } \mathbf{J} = 0 \tag{6.34}$$

$$\mathbf{B} = \mathbf{B}_{\text{ex}} + \mathbf{B}_r. \tag{6.35}$$

Here v is the velocity of the rail and, in steady state conditions, $\partial/\partial t = j\omega$.

If we know the Green's function of the Poisson equation in the air-gap, and if we assume the current density distribution in the air-gap to be known, the flux density at each point of the air-gap is obtained immediately by integration, over the whole rail, of the product of the current density and the Green's function,

$$\mathbf{B}_r(x, y, z) = \iiint_{\text{rail}} |h_1(x_1, y_1, z_1, x, y, z)\mathbf{J}_x(x_1, y_1, z_1)$$
$$+ h_2(x_1, y_1, z_1, x, y, z)\mathbf{J}_y(x_1, y_1, z_1)|\,dx_1 dy_1 dz_1. \tag{6.36}$$

In the same way, the derivative with respect to x of $\mathbf{B}_r$ is given by

$$\frac{\partial \mathbf{B}_r}{\partial x}(x, y, z) = \iiint_{\text{rail}} |k_1(x_1, y_1, z_1, x, y, z)\mathbf{J}_x(x, y, z)$$
$$+ k_2(x_2, y_2, z_2, x, y, z)\mathbf{J}_y(x, y, z)|\,dxdydz. \tag{6.37}$$

We have here two Green's functions, because the z-directed component is the sum of the field created by $\mathbf{J}_x$ (with Green's functions h_1, k_1), and $\mathbf{J}_y$ (with Green's functions h_2, k_2).

Now if we assume the value of the flux density and its derivative to be known at each position in the rail, and if $f(x_1, y_1, x, y)$ and $g(x_1, y_1, x, y)$ are Green's functions associated, respectively, with Dirichlet and Neumann boundary conditions on the rail, we can easily obtain the x and y components of the current density:

$$\mathbf{J}_x(x, y, z) = \iint_{\text{rail}} \left(-\frac{V}{\rho} \frac{\partial \mathbf{B}}{\partial x} (x_1, y_1, z_1) - j\omega/\rho \mathbf{B}(x_1, y_1, z_1) \right)$$
$$\times f(x_1, y_1, x, y) dx_1 \, dy_1. \tag{6.38}$$

$$\mathbf{J}_y(x, y, z) = \iint_{\text{rail}} \left(-\frac{V}{\rho} \frac{\partial \mathbf{B}}{\partial x} (x_1, y_1, z_1) - \frac{-j\omega}{\rho} \mathbf{B}(x_1, y_1, z_1) \right)$$
$$\times g(x, y, x_1, y_1) dx_1 dy_1. \tag{6.39}$$

For instance, if the rail is an isotropic, infinitely long strip of width L, the Green's function pick up

$$f(x, y, x_1, y_1) = +\frac{1}{4L} \left(\frac{\sin(\Pi(y + y_1 + L/2)/L)}{\cosh(\Pi(x_1 - x_1)/L) - \cos(\Pi(y + y_1 + L/2)/L)} \right. \tag{6.40}$$
$$\left. -\frac{[\sin(\Pi(y - y_1 + L/2)/L)}{\cosh(\Pi(x - x_1)/L) - \cos(\Pi(u - u_1 + L/2)/L)} \right)$$

$$g(x, y, x_1, y_1) = \frac{1}{4L} \left(\frac{\sinh(\Pi(x - x_1)/L)}{\cosh(\Pi(x - x_1)/L) - \cos(\Pi(y + y_1 + L/2)/L)} \right. \tag{6.41}$$
$$\left. -\frac{\sinh(\Pi(x - x_1)/L)}{\cosh(\Pi(x - x_1)/L) - \cos(\Pi(y - y_1 + L/2)/L)} \right).$$

Now that all the equations have been expressed in Green's function formulations, it becomes possible to discretize the rail volume and to express integrals (6.36) to (6.39) by finite sums. If we choose an integration formula, the integral (6.38) becomes, for example,

$$\mathbf{J}_x = \sum_{i,j,k} \left(-\frac{V}{\rho} \frac{\partial \mathbf{B}}{\partial x} (x_i, u_j, z_k) - j\frac{\omega}{e} \mathbf{B}(x_i \, y_i, z_i) \right) \cdot F(x_i, y_i, x, y). \tag{6.42}$$

If we arrange the values of $\mathbf{j}_x$, $\mathbf{j}_y$, $\mathbf{b}_r$, $\partial \mathbf{B}_r/\partial x$ at the nodes of the elements into vector form, we obtain the matrix formulation

$$|\mathbf{J}_x| = -\frac{V}{\sigma} |\mathbf{F}| \left| \frac{\partial \mathbf{B}_r}{\partial x} \right| - j\omega/\rho |\mathbf{F}| |\mathbf{B}_r| + F_1 |\mathbf{B}_{\text{ex}}| \tag{6.43}$$

$$|\mathbf{J}_y| = -\frac{V}{\rho} |\mathbf{G}| \left| \frac{\partial \mathbf{B}_r}{\partial x} \right| - j\omega/\rho |\mathbf{G}| |\mathbf{B}_r|. \tag{6.44}$$

Similarly, discretization of integrals (6.36) and (6.37) leads to

$$|\mathbf{B}_r| = |\mathbf{H}_1| |\mathbf{J}_x| + |\mathbf{H}_2| |\mathbf{J}_u| \tag{6.45}$$

$$\left| \frac{\partial \mathbf{B}_r}{\partial x} \right| = |\mathbf{K}_1| |\mathbf{J}_x| + |\mathbf{K}_2| |\mathbf{J}_u|. \tag{6.46}$$

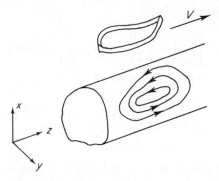

Figure 6.7   Electromagnetic suspension

The solution for vectors $\mathbf{B}_r$, $\mathbf{J}_x$, and $\mathbf{J}_y$ can be obtained by an iterative process from Equations (6.43) to (6.46) which have a large number of variables and a matrix with attractive sparsity properties. This method has been tested[6] and proved to give reasonably good results.

In the area of magnetic suspensions, and especially for high-velocity analysis, we deal with eddy currents created by the motion of a direct-current magnetic field source. The problem may be formulated as a three-dimensional Laplace equation with nonzero Neumann conditions.

Here the domain of integration is infinite in the direction of the rail, and this property allows the use of Fourier integral analysis. By the use of Fourier transforms the three-dimensional problem is transformed into a set of two-dimensional problems which may be solved by appropriate finite element techniques. A more sophisticated method using orthogonal functions has recently been used by Silvester and Nicolas and seems to give good results with less computer time than the classical Fourier analysis.

6.3.3   Nonlinear problems

The class of nonlinear problems in eddy-current analysis is a very difficult field of computation. A simple problem of this type has recently been solved by using a Galerkin projection method.[7,8] Consider the domain of Figure 6.8 on which we have to solve the equations

$$\frac{\partial}{\partial x}\left(v\frac{\partial \mathbf{A}}{\partial x}\right)+\frac{\partial}{\partial y}\left(v\frac{\partial \mathbf{A}}{\partial y}\right)=\frac{1}{\rho}\left(\frac{\partial \mathbf{A}}{\partial t}+v\frac{\partial \mathbf{A}}{\partial x}\right)\qquad y<a\qquad(6.47)$$

$$\frac{\partial}{\partial x}\left(\frac{\partial \mathbf{A}}{\partial x}\right)+\frac{\partial}{\partial y}\left(\frac{\partial \mathbf{A}}{\partial y}\right)=\frac{\mu_0}{\rho_0}\left(\frac{\partial \mathbf{A}}{\partial t}+v\frac{\partial \mathbf{A}}{\partial x}\right)\qquad y>a\qquad(6.48)$$

where v is the velocity of the slab, and with Dirichlet boundary conditions on the edges of the given domain. Equation (6.47) may be rewritten in the general

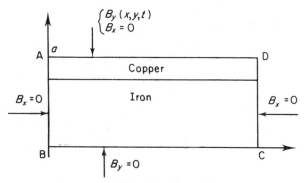

Figure 6.8 Domain of solution for two-dimensional non-linear problem

form

$$\frac{\partial \mathbf{A}}{\partial t}+v\frac{\partial \mathbf{A}}{\partial x}-\frac{\partial}{\partial x}\left(\rho v\frac{\partial \mathbf{A}}{\partial x}\right)-\frac{\partial}{\partial y}\left(\rho v\frac{\partial \mathbf{A}}{\partial y}\right)=0. \tag{6.49}$$

For use of finite element methods, we put Equation (6.49) into its Galerkin form:

$$\int_{\Omega}\left(\frac{\partial \mathbf{A}}{\partial t}+v\frac{\partial \mathbf{A}}{\partial x}\right)v(x,\ y)\mathrm{d}x\mathrm{d}y-\int_{\Omega}\left(\frac{\partial}{\partial x}\left(\rho v\frac{\partial \mathbf{A}}{\partial x}\right)+\right.$$

$$\left.\frac{\partial}{\partial x}\left(\rho v\frac{\partial \mathbf{A}}{\partial x}\right)\right)v(x,\ y)=0 \tag{6.50}$$

which must be satisfied for any continuous function $v(x,\ y)$ which vanishes outside the domain Ω. Rearranging and integrating the second term in Equation (6.50) by parts,

$$\int_{\Omega}\left(\frac{\partial \mathbf{A}}{\partial t}+v\frac{\partial \mathbf{A}}{\partial x}\right)v(x,\ y)\mathrm{d}x\mathrm{d}y+\int_{\Omega}\rho v\left(\frac{\partial \mathbf{A}}{\partial x}\frac{\partial v}{\partial x}+\frac{\partial \mathbf{A}}{\partial y}\frac{\partial v}{\partial y}\right)\mathrm{d}x\mathrm{d}y=0. \tag{6.51}$$

Discretization of this equation by means of finite elements leads us to construct a system of differential equations for the vector $\{\mathbf{A}\}$, which is the column of values of the vector potential $\mathbf{a}(x,\ y,\ t)$ at the nodes of the finite elements.

$$\mathbf{M}\frac{\mathrm{d}}{\mathrm{d}t}\{\mathbf{A}\}+\{\mathbf{N}\}\ \{\mathbf{A}\}=\mathbf{0}. \tag{6.52}$$

Matrices $\mathbf{M}$ and $\mathbf{N}$ are characteristic of the elements and of the coefficients in the above equations.

The numerical solution of (6.52) must be carried out very carefully, because this kind of system may have high stiffness, compelling us to use implicit step-by-step methods to ensure numerical stability. Nevertheless, when the computation is approached with caution, correct results are obtained without a prohibitive amount of computation.

If we shall be someday obliged to process this kind of problem in three dimensions, for three vector components with real time variation, it is certain that we shall need to use a very fast and powerful computer, because the problem will be of considerable size. We think, nevertheless, that when this kind of calamity occurs, projection methods will remain the best ones to apply, with a likelihood of solution in a finite span of time.

6.4   CONCLUSIONS

The problem of eddy current calculation is always difficult. We have attempted in the preceding pages to give some indication of solution methods for some cases specific to electric power engineering. A very large class of problems has not been investigated by nonlinear two-dimensional analysis, and the field remains open in three-dimensional cases for finding accurate methods which are not prohibitive in computer time consumption.

REFERENCES

1. Coulomb, J. L. (1976). 'Contribution à l'étude numérique des phénomènes électromagnétiques dans les machines électriques.' *Thèse*, Grenoble.
2. Coulomb, J. L., Konrad, A., Sabonnadière, J. C., and Silvester, P. (1976). Finite element analysis of steady state skin effect in a slot-embedded conductor. I.E.E.E. WPM A76-189-1.
3. Coulomb, J. L., Konrad, A., Sabonnadière, J. C., and Silvester, P., (1976). Eigenvalue analysis of steady state skin effect. 'International Conference on Numerical Methods in Electric and Magnetic fields problems.'
4. Sabonnadière, J. C. (1969). 'Contribution à l'étude des moteurs asynchrones linéaires.' *Thèse de Doctorat d'Etat*, Grenoble.
5. Nicolas, A. (1975). 'Etude des phénomènes électromagnétiques dans les moteurs asynchrones linéaires.' *Thèse*, Lyon.
6. Nicolas, A. and Sabonnadière, J. C. 'A three dimensional analysis method for linear induction machines.' I.E.E.E. WPM A76-214-7.
7. Foggia, A. (1974). Contribution à l'étude de l'induction magnétique dans les convertisseurs électromécaniques linéaires.' *Thèse*, Lyon.
8. Foggia, A., Sabonnadière, J. C., and Silvester, P., (1975). Finite element solution of saturated travelling field problem. I.E.E.E. PAS 94, May/June 1975.

Finite Elements in Electrical and Magnetic Field Problems
Edited by M. V. K. Chari and P. P. Silvester
© 1980, John Wiley & Sons Ltd.

Chapter 7

The High-Order Polynomial Finite Element Method in Electromagnetic Field Computation

Z. J. Csendes

7.1 INTRODUCTION

The application of the finite element method to electromagnetic field calculations is well established. Starting with the work of Silvester,[39] the finite element method has found increasing acceptance in many areas of electromagnetic field study and is today one of the principal methods commonly employed for the solution of these problems. However, first-order polynomial finite element methods are currently used most broadly in electromagnetic field calculations and less work is based on the more accurate and more powerful high-order polynomial finite element methods. This circumstance is unfortunate because a great deal of computer time and computer memory requirements can be saved (factors of the order of 10^3 have been reported)[7,8,40] by using high-order polynomial finite elements instead of first-order ones.

There are two principal reasons for the high popularity of the first-order finite element method compared to the high-order methods. One of these is undoubtedly due to the broad exposure afforded the first-order finite element method though the numerous successful reports in the literature.[35,53] The other reason is more fundamental: high-order finite element methods entail a considerably more complicated methodology than do their first-order counterparts. The relative complexity of the high-order finite element method has unfortunately discouraged the application of the procedure in a number of areas where it would provide significant computational advantages.

This chapter provides a review of the application of the high-order finite element method to electromagnetic field analysis. It attempts to describe in clear and simple terms the motivation and formulation of high-order polynomial finite element methods and describes both the advantages and the limitations of the procedure. Although the emphasis in this paper is to provide practical computational algorithms for the solution of field problems, a complete bibliography of high-order polynomial finite element methods in electromagnetics is included.

7.2 MATHEMATICAL BASIS

7.2.1 *Why use high-order polynomials?*

The first question invariably encountered in a discussion of high-order finite element methods is: What is the motivation for using high-order polynomial approximation functions in finite element analysis? After all, the question usually continues, first-order finite elements are so effective and work so well that there is no need for more sophisticated higher-order methods, or is there?

The response to this question is yes, there is a need for high-order finite elements and the need arises from a desire to increase the accuracy or to decrease the cost of a finite element solution. Although first-order finite elements do work well, given a fixed amount of computer time the accuracy of any first-order finite element solution can be improved by using high-order polynomial finite elements instead. And the improvement in accuracy can be dramatic. A fourth-order finite element mesh of one hundred nodes (independent variables) is equivalent in accuracy to a first-order mesh of many thousands of nodes.

An intuitive understanding of the nature of high-order polynomial finite element approximation can be gained by considering the following simple example. Suppose that the function $\sin(x)$ is to be approximated in the interval $x = [0, \pi/2]$ by zeroth, first, and second-order polynomials in such a way that the number of independent parameters in each approximation is the same. Using a simple point-matching approximation at the three points $x = 0$, $\pi/4$, $\pi/2$, for example, the three approximations in Figure 7.1 result. Clearly, the two-element linear approximation is much more accurate than the three-element constant approximation, and the single-element quadratic approximation is much more accurate than the two-element linear approximation. (More precisely, the L_2 norms $\eta = \| y - y_{\text{app}} \|$ in the three cases are (a) $\eta = 0.040$, (b) $\eta = 0.0021$, (c) $\eta = 0.00037$.) In terms of a three-point finite element approximation of a differential equation with the solution $y = \sin(x)$, this result implies that a great increase in solution accuracy (of the order of 10^2) can be obtained by using second-order finite elements instead of zeroth-order elements, even though a three by three matrix equation must be solved in each case.

As is the case in changing from zeroth to first, or from first to second order polynomial finite elements, discretization error is reduced for a given number of parameters in using each consecutive higher order polynomial finite element. However, at about polynomial orders of seven or eight, round-off error in polynomial construction becomes significant and prevents efficient higher-order polynomial element use.

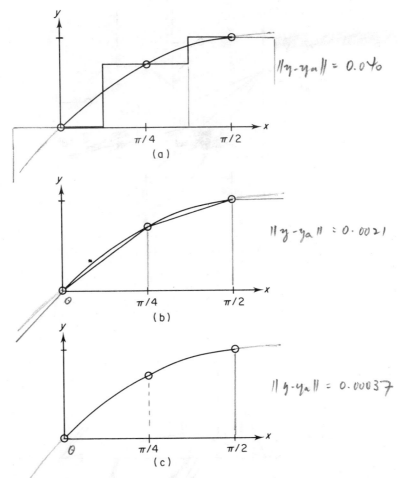

Figure 7.1 Three-point approximations of $y = \sin(x)$ in the interval $x = [0, \pi/2]$. (a) Three element constant function (zeroth-order polynomial) approximation; (b) two element first-order polynomial approximation; and (c) one element second-order polynomial approximation

7.2.2 Triangular interpolation polynomials

The basic theory of high-order polynomial interpolation over triangles was developed independently but simultaneously by Silvester[41,43] and by Irons[22], using two fundamentally different approaches to the problem. As described by Irons, an algorithm for generating high-order interpolation polynomials over

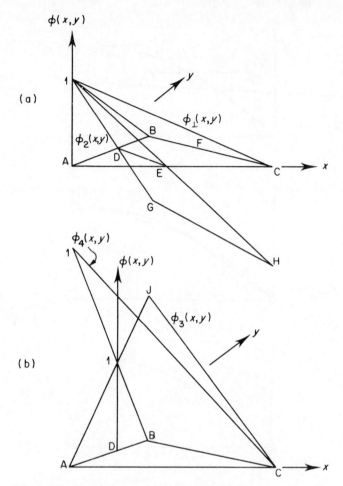

Figure 7.2 Generation of quadratic interpolation polynomials on a triangle. (a) Vertex interpolation function; (b) side-point interpolation function

triangular regions is obtained by taking the product of several different first-order polynomials in the same triangle. For example, Figure 7.2(a) presents a picture of a triangle A–B–C over which two first-order polynomial shape functions $\phi_1(x, y)$ and $\phi_2(x, y)$ are defined. These two functions have unit value $\phi = 1$ at vertex A but $\phi_1(x, y)$ intersects the triangle A–B–C along the line B–C (i.e. $\phi_1(x, y) = 0$ on B–C) and $\phi_2(x, y)$ intersects triangle A–B–C along the line D–E. The product of the two linear functions $\phi(x, y) = \phi_1(x, y)\phi_2(x, y)$ is therefore a quadratic polynomial in x and y which is unity at vertex A and is zero along the lines D–E and B–C.

In much the same way, a quadratic polynomial which has unit value at point D and is zero along triangle sides A–C and B–C can be constructed as the product of two first-order polynomials $\phi_3(x, y)$ and $\phi_4(x, y)$ having a unit value at point D and passing through zero along the lines A–C and B–C respectively, as indicated in Figure 7.1(b). With a procedure for generating interpolatory polynomials at points A and D in hand, a similar reasoning can be applied to produce quadratic polynomials which interpolate at each of the remaining triangle points B,C,E, and F.

7.2.3 Silvester's approach

As developed by Irons, the procedure for generating high-order interpolation polynomials in triangles is general but cumbersome. This is because the first-order components in the product polynomial $\phi(x, y)$ must be evaluated and properly normalized in each application. Fortunately, in the alternate analysis developed by Silvester[41] a very simple and elegant algebraic representation of high-order interpolation polynomials is obtained in the special case in which the interpolation nodes form a regular, equi-spaced grid on the triangle. (In Irons' method, the interpolation nodes D and E may be placed anywhere on the two sides A–B and A–C, respectively.) The heart of Silvester's method is the polynomial

$$P_m^{(N)}(\zeta) = \left(\frac{N\zeta - m + 1}{m}\right) P_{m-1}^{(N)}(\zeta) \qquad \text{if } m > 0$$

$$P_0^{(N)}(\zeta) = 1$$

$$P_m^{(N)}(\zeta) = 0 \qquad \text{if } m < 0$$

(7.9)

Since for $m > 0$, $P_m^{(N)}(\zeta)$ equals zero at $\zeta = k/N$, $k = 0, 1, 2, \ldots, m-1$, the polynomials $(N\zeta - k + 1)/k$ are seen to provide exactly the correct first-order polynomial factors required for the construction of the high-order elements. Defining triangle homogeneous coordinates ζ_1, ζ_2, and ζ_3 as the ratios of the distances between a point inside a triangle and its three sides to the corresponding altitudes of the triangle,[32] the product polynomial

$$\alpha_{ijk}^{(N)}(x, y) = P_i^{(N)}(\zeta_1) P_j^{(N)}(\zeta_2) P_k^{(N)}(\zeta_3) \qquad i + j + k = N$$

(7.2)

therefore yields an interpolation polynomial which is unity at the node $\zeta_1 = i/N$, $\zeta_2 = j/N$, $\zeta_3 = k/N$ but equals zero at all other nodes $\zeta_1 = p/N$, $\zeta_2 = q/N$, $\zeta_3 = r/N$ for (p, q, r) integers not equal to (i, j, k) (see Figure 7.3).

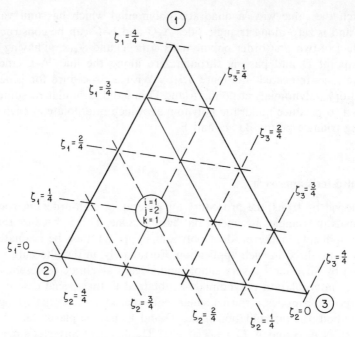

Figure 7.3 Graphic interpolation of the properties of the interpolation polynomial $\alpha_{12,1}^{(4)}$. This polynomial is zero along the lines $\zeta_1 = 0,\; \zeta_2 = 0,\; \zeta_2 = 1/4,\;$ and $\zeta_3 = 0$

7.2.4 Matrix notation

An arbitrary function $\phi(x, y)$ can be approximated in a triangle in terms of the interpolation polynomials $\alpha\ (x, y)$ by the sum

$$\phi(x, y) = \sum_{\substack{ijk = 0 \\ i+j+k=N}}^{N} \phi_{ijk} \alpha_{ijk}^{(N)}(x, y) \tag{7.3}$$

where the ϕ_{ijk} are arbitrary coefficients to be determined. In terms of the polynomials $P_M^{(N)}(\zeta_i)$, equation (7.3) can be written in the convenient matrix form

$$\phi(x, y) = \mathbf{P}^{(N)}(\zeta_2) \mathbf{H}^{(N)}(\zeta_1) \mathbf{P}^{(N)}(\zeta_3)^T \tag{7.4}$$

where $\mathbf{P}^{(N)}(\zeta_i)$ is the row vector

$$\mathbf{P}^{(N)}(\zeta_i) = [P_0^{(N)}(\zeta_i)\, P_1^{(N)}(\zeta_i) \ldots P_N^{(N)}(\zeta_i)] \tag{7.5}$$

and $\mathbf{H}^{(N)}(\zeta_1)$ is an $(N+1)$ by $(N+1)$ matrix with elements

$$H_{ij}^{(H)}(\zeta_1) = P_{n-i-j+2}^{(N)}(\zeta_1) \phi_{N-i-j+2,i-1,j-1}. \tag{7.6}$$

Note that all of the entries in the lower right-hand triangle of the matrix $\boldsymbol{H}^{(N)}(\zeta_1)$ are zero. For example, for $N=2$

$$^{(N)}(\zeta_1)=\begin{bmatrix} P_2(\zeta_1)\phi_{200} & P_1(\zeta_1)\phi_{101} & P_0(\zeta_1)\phi_{002} \\ P_1(\zeta_1)\phi_{110} & P_0(\zeta_1)\phi_{011} & 0 \\ P_0(\zeta_1)\phi_{020} & 0 & 0 \end{bmatrix} \tag{7.7}$$

Equation (7.4) may be converted to standard form by applying the operation vec to both sides (vec $\mathbf{A}=[A_1 A_2 \ldots A_n]^T$ where A_i is the ith column of a matrix $\mathbf{A}$)[37]; this gives

$$\phi(x, y) = (\mathbf{P}^{(N)}(\zeta_3) \times \mathbf{P}^{(N)}(\zeta_2))\boldsymbol{R}^{(N)}(\zeta_1)\boldsymbol{\phi}^{(N)} \tag{7.8}$$

where $\times$ denotes the matrix kronecker product $\mathbf{A} \times \mathbf{B} = [a_{ij}\mathbf{B}]$[37], $\mathbf{R}^{(N)}(\zeta_1)$ is a diagonal matrix with elements

$$R_{ii}^{(N)}(\zeta_1) = P_k^{(N}(\zeta_1) \tag{7.9}$$

where $k = N - \text{mod}\,(i-1, N+1) - [(i-1)/(N+1)]$ and $\boldsymbol{\phi}$ is the column vector

$$\boldsymbol{\phi}^{(N)} = \begin{bmatrix} \phi_{N,0,0} \\ \phi_{N-1,1,0} \\ \cdot \\ \cdot \\ \cdot \\ \phi_{0,N,0} \\ \hline \phi_{N-1,0,1} \\ \cdot \\ \cdot \\ \cdot \\ \phi_{0,N-1,1} \\ 0 \\ \hline \phi_{N-2,0,2} \\ \cdot \\ \cdot \\ \cdot \\ \phi_{0,N-2,2} \\ 0 \\ 0 \\ \hline \cdot \\ \cdot \\ \cdot \end{bmatrix} \tag{7.10}$$

(In the above, mod represents the modulo function and [/] denotes integer division.) Equation (7.8) may also be written as

$$\phi(x,\ y)=\boldsymbol{\alpha}^{(N)}(x,\ y)\phi^{(N)} \tag{7.11}$$

where $\boldsymbol{\alpha}^{(N)}(x,\ y)$ is the row vector

$$\boldsymbol{\alpha}^{(N)}(x,\ y)=(\mathbf{P}^{(N)}(\zeta_3)\times\mathbf{P}^{(N)}(\zeta_2))\,\boldsymbol{R}^{(N)}(\zeta_1). \tag{7.12}$$

7.2.5   Finite element matrices

In the finite element method, a number of matrices involving integrals of products of the interpolation polynomials $\boldsymbol{\alpha}^{(N)}(x,\ y)$ need to be evaluated. For example, the discretization of Poisson's equation results in matrix integrals of the form

$$\boldsymbol{S}=\int_A \nabla\boldsymbol{\alpha}^T(x,\ y).\nabla\boldsymbol{\alpha}(x,\ y)\mathrm{d}x\mathrm{d}y \tag{7.13}$$

$$\boldsymbol{T}=\int_A \boldsymbol{\alpha}^T(x,\ y)\,\boldsymbol{\alpha}(x,\ y)\mathrm{d}x\mathrm{d}y$$

where A is the area of the triangle. Silvester[41] has shown that the matrix $\boldsymbol{S}$ may be written in the form

$$\boldsymbol{S}=\sum_{i=1}^{3}\cot\ \theta_i\boldsymbol{Q}^{(i)} \tag{7.14}$$

where θ_i is the angle subtended by the triangle at vertex i and

$$\boldsymbol{Q}^{(i)}=\int (G^{(i)}\boldsymbol{\alpha}(x,\ y))^T(G^{(i)}\boldsymbol{\alpha}(x,\ y))\mathrm{d}x\mathrm{d}y \tag{7.15}$$

where $G^{(i)}$ is the operator

$$G^{(i)}=\frac{\partial}{\partial\zeta_{\mathrm{mod}(i+1,3)}}-\frac{\partial}{\partial\zeta_{\mathrm{mod}(i+2,3)}}. \tag{7.16}$$

Substituting Equation (7.12) into Equation (7.13) yields for $\boldsymbol{T}$

$$\boldsymbol{T}=A\int_{\zeta_1=0}^{1}\boldsymbol{R}(\zeta_1)\int_{\zeta_2=0}^{1-\zeta_1}\int_{\zeta_3=0}^{1-\zeta_1-\zeta_2}\mathbf{P}(\zeta_3)T_\mathrm{P}(\zeta_3)\mathrm{d}\zeta_3\times\mathbf{P}^T(\zeta_2)\mathbf{P}(\zeta_2)\mathrm{d}\zeta_2\,\boldsymbol{R}(\zeta_1)\mathrm{d}\zeta_1 \tag{7.17}$$

and a similar expression is obtained for $\boldsymbol{Q}^{(i)}$. Both $\boldsymbol{T}$ and $\boldsymbol{Q}^{(i)}$ are independent of the triangle shape and $\boldsymbol{Q}^{(i)}$ is independent of the triangle area as well.

7.3 COMPUTATIONAL IMPLEMENTATION

7.3.1 Integration procedure

The elements of the finite element matrices in Equations (7.15) and (7.17) are integrals of the products of many different high-order polynomials. As such, these integrals are difficult (if not impossible) to evaluate by hand and computer methods must be employed. One possible procedure for evaluating the finite element integrals is the application of Gaussian quadrature formulas and this method has a wide following in both the engineering and the mathematics communities[53,35,49,36]. However, numerical integration is a costly procedure, and can result in additional computational error if improperly performed. For these reasons, Silvester[41] has proposed a method for evaluating the finite element matrices exactly in all cases and then tabulating the results in terms of basic triangle properties such as the triangle area and the triangle vertex angles.

In the original work, Silvester[41] developed his own computer programs to evaluate the finite element matrices T and $Q^{(i)}$ for polynomial orders $N = 1$ to 4. Subsequent work in the field has, however, been mainly performed with commercially available symbol manipulation computer languages. There are three principal symbol manipulation computer languages available today: FORMAC, developed by IBM for use with IBM series 360 and 370 computers; ALTRAN, developed at Bell Laboratories principally for use with Honeywell computers; and MACSYMA, developed at the Massachusetts Institute of Technology on the MIT Mathlab computer[52,2]. These symbol manipulation computer languages have made it astonishingly easy to evaluate the complicated algebraic expressions encountered in finite element discretization.

In addition to the matrices S and T, corresponding to the Laplacian operator and the interpolation polynomial metric matrix, a number of other finite element matrix operators have been evaluated and tabulated in recent years. One of these is the anti-symmetric matrix

$$U = \int \left(\frac{\partial \boldsymbol{\alpha}^T(x, y)}{\partial x} \frac{\partial \boldsymbol{\alpha}(x, y)}{\partial y} - \frac{\partial \boldsymbol{\alpha}^T(x, y)}{\partial y} \frac{\partial \boldsymbol{\alpha}(x, y)}{\partial x} \right) dx dy \qquad (7.18)$$

which was first evaluated independently by Daly[15] and by Csendes[6]. Two others are the symmetric matrix

$$V = \frac{1}{2A} \int \left(\boldsymbol{\alpha}^T(x, y) \frac{\partial \boldsymbol{\alpha}(x, y)}{\partial x} + \frac{\partial \boldsymbol{\alpha}^T(x, y)}{\partial x} \boldsymbol{\alpha}(x, y) \right) dx dy \qquad (7.19)$$

and the anti-symmetric matrix[50,51]

$$W = \frac{1}{2A} \int \left(\boldsymbol{\alpha}^T(x, y) \frac{\partial \boldsymbol{\alpha}(x, y)}{\partial x} - \frac{\partial \boldsymbol{\alpha}^T(x, y)}{\partial x} \boldsymbol{\alpha}(x, y) \right) dx dy. \qquad (7.20)$$

Konrad[25] has evaluated all of the above matrices and their rotationally symmetric counterparts (in which the integrals are weighted by the radial coordinate *r*) for polynomials of orders one to six inclusive. Konrad has also published a Fortran computer program for generating data statements containing all of the above matrix elements.[25]

7.3.2 Programming considerations

High-order polynomial finite element methods introduce several programming considerations not encountered in first-order finite element computer programs. First of all, as mentioned earlier in the paper, a few high-order finite elements are more accurate than many first-order elements. Therefore, high-order finite element triangulations have fewer elements and appear to have a different form than do first-order triangulations. In a high-order finite element triangulation, very few high-order elements are used in the interior section of a problem to model large homogeneous sections, with possibly many small, low-order elements adjacent to a boundary of complicated shape. A typical high-order finite element triangulation is presented in Figure 7.4; note the use of mixed-order finite elements in this problem.

Another important consideration in a high-order finite element computer

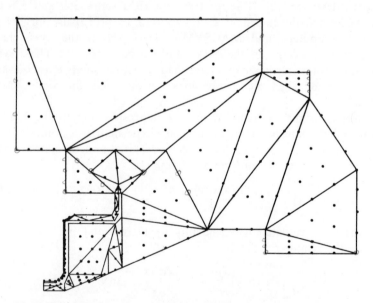

Figure 7.4 A high-order finite element triangulation of the end region of a generator for electrostatic analysis. This triangulation consists of 21 fourth order triangles and 40 second order triangles; the total number of interpolation nodes is 322

program is the coefficient matrix structure generated. Since the number of nodes in a single high-order finite element is much greater than the three nodes in a single first-order finite element, high-order finite element coefficient matrices are considerably less sparse than first-order coefficient matrices. This implies that it is far more difficult to exploit the bandedness of the high-order coefficient matrices and virtually impossible to obtain a good node ordering by using an intuitive procedure. A very important development in this regard is the study by George[21] which showed that a very efficient finite element node numbering is generated by reversing the ordering produced by the Cuthill–McKee algorithm, developed for use in electrical network analysis.[5] At the present time, many efficient high-order polynomial finite element computer programs employ the reverse Cuthill–McKee algorithm for node ordering, followed by a sparse Gaussian elimination routine for matrix solution.

There are several published high-order polynomial finite element computer programs available in the literature. The first of these is the original high-order finite element program developed by Silvester specifically for the solution of waveguide problems.[42] Next, there is a high-order finite element computer program for the solution of dielectric loaded waveguides.[12] Thirdly, there is a high-order finite element computer program for the solution of Laplace's, Poisson's, and Helmholtz's equations.[27] Fourthly, there are two finite element computer programs for the solution of axisymmetric scalar field problems.[28],[30] Each of these programs is published with a description of the program operation and usage and also a complete listing of the computer code.

7.4 APPLICATIONS

The applications of the high-order finite element method in electromagnetics are in many cases the same as the applications of the first-order finite element method, except that no pre-computed matrix high-order solutions of nonlinear magnetics problems have appeared. The three principal areas of interest are electrostatics, microwaves, and magnetic field analysis.

7.4.1 Waveguide analysis

The first reported application of the high-order polynomial finite element method was by Silvester[40] for the analysis of homogeneous waveguide problems. The axial components of the electric and the magnetic fields in a waveguide are governed by the Helmholtz equation

$$\mathbf{V}^2 \phi = k^2 \phi \tag{7.21}$$

subject to the homogeneous Dirichlet and Neuman boundary conditions. In the high-order finite element method, Equation (7.21) is discretized to yield the matrix equation

$$S\phi = k^2\, T\phi \qquad (7.22)$$

In a typical case, the percent error in the cutoff wavelength of an H_{21} mode in a 2.25:1 rectangular waveguide is approximately 3% using an 80 node first-order finite element model but only 0.015% using an 80 node fourth-order finite element model.[40] In fact, a simple 25 node fourth-order finite element model produced better results, with 1.5% error, than did the 80 node first-order solution.

Subsequent applications of high-order finite elements to homogeneous waveguide problems have been published.[4,31,47] Daly[18] has developed a high-order finite element method for the solution of polar geometry waveguides in which the waveguide boundaries coincide with coordinate surfaces in circular polar coordinates.

The finite element solution of the Helmholtz equation (7.21) has also been useful in the study of microwave planar network structures.[44] In this analysis, the eigenvalues and eigenfunctions of the planar network geometry are obtained by means of the high-order polynomial finite element method (Equation (7.22)) and then these modal solution values are related to the admittance matrix of the N-port microwave network.

7.4.2   Inhomogeneous structures

A second application of the high-order polynomial finite element method is the study of dielectric loaded waveguides.[11,13] In this formulation, the axial components of the electric and magnetic fields are coupled by the matrix U in Equation (7.18) and a matrix equation of the form

$$\frac{1}{\epsilon\mu - \delta^2}\begin{bmatrix} \epsilon S & -\delta U \\ -\delta U & \mu S \end{bmatrix}\begin{bmatrix} E_z \\ H_z \end{bmatrix} = \omega^2 c\begin{bmatrix} \epsilon T & 0 \\ 0 & \mu T \end{bmatrix}\begin{bmatrix} E_z \\ H_z \end{bmatrix} \qquad (7.23)$$

results, where $\delta = \beta/\omega$. The method works well for low values of relative phase velocity δ but encounters serious difficulty at high values of relative phase velocity. The cause of this difficulty is the indefinite nature of the finite element coefficient matrix in high δ cases, which results in extraneous, nonphysical modes in the dielectric loaded waveguide eigenvalue spectrum.

Some typical examples of electric and magnetic field eigenfunction contour plots produced by the high-order finite element method are presented in Figure 7.5. All of these contour plots were produced with a second-order finite element field plotting routine,[14] although various orders of finite element approximation functions were used in the solution process.

Independently of this, Daly[16] developed and applied a very similar procedure for the study of wave propagation on a microstrip placed inside a conducting box. Recently, McAulay[33] extended the same finite element pro-

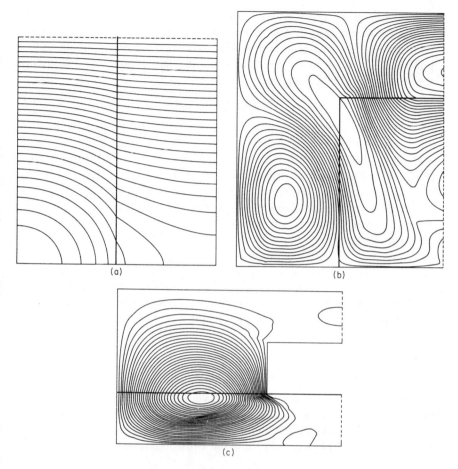

Figure 7.5 Examples of dielectric loaded waveguide eigenfunction contour plots (a) H_z field of the dominant mode at cutoff; (b) E^z field of the second mode at $\delta = 1.5$; (c) E^z field of the dominant waveguide mode at $\delta = 0.9$

cedure to the study of dissipative electromagnetic surface waveguides. Unfortunately, although many microstrip and surface wave propagation problems occur in unbounded space, both of these analyses are limited to the study of wave propagation in finite, bounded regions.

The only reported solution of unbounded microwave propagation problems using the high-order finite element method is by McDonald and Wexler.[34,38] In the procedure they developed, the region of interest is surrounded by two arbitrary concentric boundaries on which the electromagnetic field is related by both the finite element equations and a Hankel function Green's function.

The electromagnetic field is solved only within the outer boundary but has the same appearance within that boundary as if it were determined in all space.

In order to avoid the nonphysical solutions obtained by the application of Equation (7.23), Konrad[25] reformulated the dielectric loaded waveguide problem in terms of the curlcurl equation

$$\nabla \times \hat{\mathbf{p}}\ \nabla \times \psi = \omega^2 \hat{\mathbf{q}}\ \psi \qquad (7.24)$$

where $\hat{\mathbf{p}}$ and $\hat{\mathbf{q}}$ are Hermitian tensors corresponding to the permeability and permittivity of the medium, and the three-component vector ψ is either the electric field intensity vector $\mathbf{E}$ or the magnetic field intensity vector $\mathbf{H}$. This high-order finite element formulation has the advantage that the numerical difficulties inherent in Equation (7.23) are avoided and that anisotropic waveguide loading can be modelled. However, the procedure requires that somewhat larger matrix eigenvalue problems be solved than with the two-component vector finite element formulation and is therefore less efficient in execution.

A completely different approach to the numerical solution of microstrip problems is the high-order finite element current density representation.[23] In this approach, a nonlinear integral equation is provided for the current density distribution on the microstrip conductor and this integral equation is solved by finite element approximation.

7.4.3   Electrostatic field analysis

The analysis of electrostatic field problems via the high-order polynomial finite element method is very similar in form to waveguide analysis, except that in this case a deterministic problem is solved and not an eigenvalue problem. The governing equation of electrostatics is the Poisson equation

$$\nabla^2 \phi = -\rho/\epsilon \qquad (7.25)$$

which is discretized to yield the matrix equation

$$S\phi = T\rho. \qquad (7.26)$$

These equations are to be compared with Equations (7.21) and (7.22) appearing in waveguide analysis. The similarity of the above equations makes it possible for both Equations (7.22) and (7.26) to be formed by the sample program segment but, of course, two different matrix solution algorithms are required for processing.[17,27,24,29]

An example of a typical high-order finite element solution of an electrostatic field problem is provided by the example given in Figure 7.4. In this problem, which occurs during high-potential testing of the generator armature winding, the armature winding is raised to a potential of 55 KV while the generator body, rotor, and core remain at ground potential. A contour plot of the equi-

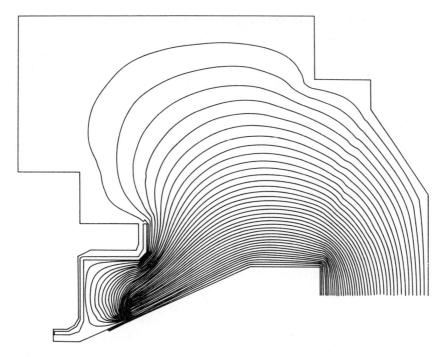

Figure 7.6 Electrostatic potential contours for the problem in Figure 7.4. The solution presented here is axisymmetric and required approximately 36 seconds CPU time for solution on a Honeywell 6080 computer

potential lines produced by high-order finite element analysis for this problem is shown in Figure 7.6.

Due to the variational nature of the finite element approximation procedure, global quantities such as energy and capacitance can be computed directly from the solution potential coefficients and the finite element matrices S and T.[48,54]

A high-order finite element method for the solution of unbounded electrostatics field problems was first published by Silvester and Hsieh.[46] In contrast to the exterior-field formulation[34] described earlier (which is applicable not only to wave propagation problems but also to electrostatic problems), Silvester and Hsieh developed a method by which the energy functional contribution from all space is considered, but in which the field is computed only within the confines of an arbitrary boundary. The method works well and is highly accurate; however, the procedure is quite difficult to program because a large number of singular integrals must be evaluated numerically. In order to avoid these computations, Csendes[9] has proposed a geometric approximation by which the exterior field boundary integrals can be evaluated analytically and tabulated, using a similar procedure to that employed with the interior field finite element matrices S and T.

In recent years, the finite element method has also been used to determine the electric potential in semiconductor devices.[1,19] These papers, however, make use of only first- and second-order triangular finite elements.

4.4 Magnetic field calculations

A complete review of finite element methods for magnetic field calculations with specific reference to electrical machine problems is given by Chari.[3] Here, only those computations which involve high-order polynomial finite element solutions will be described.

A high-order polynomial finite element analysis of the diffusion equation

$$\mathbf{V}^2\phi = -j\omega\mu\sigma\phi \tag{7.27}$$

has been presented[45] with regard to magnetotelluric prospecting calculations. The finite element matrix representation of this equation is

$$S\boldsymbol{\phi} = j\omega\mu\sigma T\phi^* \tag{7.28}$$

which again bears a close resemblance to Equation (7.22) encountered in waveguide analysis. A second method of solution of Equation (7.27) using model analysis has also been published.[26] This solution is analogous to the eigenvalue technique developed by Silvester[44] for the solution of microwave planar network problems described previously.

Another type of problem in magnetic field analysis to which the high-order polynomial finite element method has been applied is the calculation of magnetic fields involving moving conductors. This analysis requires the construction of an unsymmetric matrix from a linear combination of the matrices V and W given in Equations (7.19) and (7.20). Details of the computational procedure are available for rectilinear motion[20] and for the case of rotating machinery.[10]

One additional paper for the solution of magnetic field problems by the high-order finite element method has appeared in the literature, on the analysis of rotationally symmetric transformer field leakage phenomena.[48] Of particular interest in this paper is the construction of matrices which replace the process of differentiation on high-order polynomial finite elements. A procedure for discretizing arbitrary partial differential equations using a similar procedure has also been reported.[7]

7.5 CONCLUSIONS

The high-order polynomial finite element method has become an established procedure for the solution of electromagnetic field problems. The method is currently being used to analyse a wide variety of field problems—from waveguides and microwave planar networks to transformers and large synchronous machinery—and the list of successful applications of the method is likely to continue to grow in the future.

Undoubtedly, the acceptance of the high-order polynomial finite element method in electromagnetic field computation is due to the very high accuracy of the technique, coupled with its modest computer memory requirements and low operating cost. The general philosophy of the high-order polynomial finite element method is to increase the accuracy of the approximating functions while at the same time limiting the number of independent parameters in the solution process. This has led in the high-order polynomial finite element method to smaller, more compact but, at the same time, more dense coefficient matrices than are obtained with the first-order finite element method. The resulting small, dense coefficient matrix equations can be solved much more economically than the large, very sparse systems obtained with first-order elements.

One of the cardinal principles of the high-order finite element method described in this chapter is the use of analytically pre-computed integrals for finite element matrix generation. The use of pre-computed integrals in the finite element process reduces the otherwise lengthy and complicated process of finite element coefficient matrix evaluation to a simple and efficient bookkeeping operation. In this process the global values in the finite element coefficient matrix are obtained by simply reading the appropriate entries from a list of precomputed integrals.

ACKNOWLEDGEMENT

The author would like to thank the General Electric Company for permission to publish this contribution.

REFERENCES

1. Barnes, J. J., and Lomas, R. J. (1974). 'Two-dimensional finite element simulation of semiconductor devices', *Electronics Letters*, **10**, 341–342.
2. Brown, W. S. (1973). *Altran Users Manual*, Bell Telephone Laboratories, Murray Hill.
3. Chari, M. V. K. (1976). 'Finite element solution of magnetic and electric field problems in electrical machines and devices'. Published in M.V.K. Chari and P. Silvester, *Finite Elements in Electric and Magnetic Field Problems*, John Wiley, London, 1979.
4. Chen, N. H., Tsandoulas, G. N., and Willwerth, F. G. (1974). 'Modal characteristics of quadruple-ridged circular and square waveguides', *IEEE Trans.*, **MTT-22**, 801–804.
5. Cuthill, E. H. and McKee, J. M. (1969). 'Reducing the bandwidth of sparse symmetric matrices', *Proc. 24th Conference ACM*, pp. 157–172.

6. Csendes, Z. J. (1970). 'Solution of Dielectric Loaded Waveguides by Finite Element Methods', *M. Eng. Thesis*, McGill University.
7. Csendes, Z. J. (1975a). 'A Fortran program to generate finite difference formulas', *Int. J. Numerical Methods in Engineering*, **9**, 581–599.
8. Csendes, Z. J. (1975b). 'A finite element method for the general solution of ordinary differential equations', *Int. J. Numerical Methods in Engineering*, 9, 551–561.
9. Csendes, Z. J. (1976). 'A note on the finite element solution of exterior-field problems', *IEEE Trans.*, **MTT-24**, No. 6.
10. Csendes, Z. J. and Chari, M. V. K. (1977). 'Finite element analysis of eddy current effects in rotating electric machines', presented at the IEEE PES Summer Meeting, Mexico City, Mex., paper number A 77 615–8.
11. Csendes, Z. J. and Silvester, P. (1970). 'Numerical solution of dielectric loaded waveguides: I. Finite–element analysis', *IEEE Trans.*, **MTT-18**, No. 12, 1124–1131.
12. Csendes, Z. J. and Silvester, P. (1971a). 'Dielectric loaded waveguide analysis program', *IEEE Trans.*, **MTT-19**, 789.
13. Csendes, Z. J. and Silvester, P. (1971b). 'Numerical solution of dielectric loaded waveguides: II. Modal approximation technique', *IEEE Trans.*, **MTT-19**, No. 6, 504–509.
14. Csendes, Z. J. and Silvester, P. (1972). 'Finplt: a finite-element field plotting program', *IEEE Trans.*, **MTT-20**, No. 4, 294–295.
15. Daly, P. (1969). 'Finite-element coupling matrices', *Electronics Letters*, **5**, No. 24, 613–615.
16. Daly, P. (1971). 'Hybrid-mode analysis of microstrip by finite element methods', *IEEE Trans.*, **MTT-19**, No. 1, 19–25.
17. Daly, P. (1973). 'Singularities in transmission lines'. In J. R. Whiteman (Ed), *The Mathematics of Finite Elements and Applications*, Academic Press, New York. pp. 337–350.
18. Daly, P. (1974). 'Polar geometry waveguides by finite element methods', *IEEE Trans.*, **MTT-22**, 202–209.
19. De Mey, G. (1975). 'The finite element method for potential calculations in a Hall plate', *The Radio and Electronic Engineer*, **45**, 472–474.
20. Foggia, A., Sabonnadière, J. C., and Silvester, P. (1975). 'Finite element solution of saturated travelling magnetic field problems', *IEEE Trans.*, **PAS-94**, No. 3.
21. George, J. A. (1971). 'Computer Implementation of the Finite Element Method', *Ph.D. Thesis*, Stanford University, AD 726 171.
22. Irons, B. M. (1969). 'Economical computer techniques for numerically integrating finite elements', *Int. J. Numerical Methods in Eng.*, **1**, 201–204.
23. Jansen, R. (1975). 'Finite element current density representation in the numerical solution of microstrip problems', *AEU*, **29**, 477–480.
24. Konrad, A. (1973). 'Linear accelerator cavity field calculation by the finite element method', *IEEE Trans. Nucl. Sci.*, **NS-20**, No. 1, Feb. 1973, 802–808.
25. Konrad, A. (1974). 'Triangular Finite Elements for Vector Fields in Electromagnetics', *Ph.D. thesis*, McGill University.
26. Konrad, A., Coulomb, J. L., Sabonnadière, J. C., and Silvester, P. (1976). 'Finite element analysis of steady-state skin effect in a slot-embedded conductor', *IEEE Winter Meeting, New York*, Paper No. A 76 189–1.
27. Konrad, A. and Silvester, P. (1971). 'Scalar finite element program package for two-dimensional field problems', *IEEE Trans.*, **MTT-19**, No. 12, 952–954.
28. Konrad, A. and Silvester, P. (1973a). 'Triangular finite elements for the generalized Bessel equation of order m', *Int. J. Num. Meth. Eng.*, 7, No. 1, 43–55.
29. Konrad, A. and Silvester, P. (1973b). 'A finite element program package for axisymmetric scalar field problems', *Comp. Phys. Comm.*, **5**, 437–455.

30. Konrad, A. and Silvester, P. (1975). 'A finite element program package for axisymmetric vector field problems', *Comp. Phys. Comm.*, **9**, 193–204.
31. Lagasse, P. and van Bladel, J. (1972). 'Square and rectangular waveguides with rounded corners', *IEEE Trans.*, **MTT-20**, 331–337
32. Maxwell, E. A. (1946). *General Homogeneous Coordinates*, University Press, Cambridge.
33. McAulay, A. D. (1974). 'Detection of Track Guided Ground Vehicles Using the Track as an Electromagnetic Waveguide', *Ph.D. Thesis*, Carnegie Mellon University.
34. McDonald, B. H. and Wexler, A. (1972). 'Finite-element solution of unbounded field problems', *IEEE Trans.*, **MTT-20**, 841–847.
35. Oden, J. T. (1972). *Finite Elements of Nonlinear Continua*, McGraw-Hill, New York.
36. Prenter, P. M. (1975). *Splines and Variational Methods*, Wiley, New York.
37. Rao, C. R. and Mitra, S. K. (1971). *Generalized Inverse of Matrices and its Applications*, Wiley, New York.
38. Richards, D. J. and Wexler, A. (1972). 'Finite element solutions within curved boundaries', *IEEE Trans.*, **MTT-20**, 650–657.
39. Silvester, P. (1968). 'Finite element solution of homogeneous waveguide problems', 1968 *URSI Symp. Electromagnetic Waves*. Published in *Alta Frequenza*, **38** (1969), 313–317.
40. Silvester, P. (1969a). 'A general high-order finite element waveguide analysis program', *IEEE Trans. Microwave Theory Tech.*, **MTT-17**, 204–210.
41. Silvester, P. (1969b). High-order polynomial triangular finite elements for potential problems', *Int. J. Engineering Science*, **7**, 849–861.
42. Silvester, P. (1969c). 'High-order finite element waveguide analysis', *IEEE Trans.*, **MTT-17**, No. 8, 651.
43. Silvester, P. (1970). 'Symmetric quadrature formulae for simplexes', *Math. Comp.*, **24**, 95–100.
44. Silvester, P. (1973). 'Finite element analysis of planar microwave networks', *IEEE Trans.*, **MTT-21**, 104–108.
45. Silvester, P. and Haslam, C. R. S. (1972). 'Magnetotelluric modelling by the finite element method', *Geophysical Prospecting*, **20**, 872–891.
46. Silvester, P. and Hsieh, M. S. (1971). 'Finite element solution of two-dimensional exterior field problems', *Proc. IEE, 118*, No. 12, 1743–1747.
47. Silvester, P. and Konrad, A. (1973a). 'Axisymmetric triangular finite elements for the scalar Helmholtz equation', *Int. J. Num. Meth. Eng.*, **5**, 481–497.
48. Silvester, P. and Konrad, A. (1973b). 'Analysis of transformer leakage phenomena by high-order finite elements', *IEEE Trans.*, **PAS-92**, No. 6.
49. Strang, G. and Fix, G, J, (1973). *An Analysis of the Finite Element Method*, Prentice-Hall, Englewood Cliffs, N. J.
50. Stone, G. O. (1972). 'Coupling matrices for high-order finite element analysis of acoustic-wave propagation', *Electronics Letters*, **8**, No. 18, 466–468.
51. Stone, G. O. (1973). 'High-Order finite elements for inhomogeneous acoustic guiding structures', *IEEE Trans.*, **MTT-21**, No. 8, 538–542.
52. Xenakis, J. (1969). *PL/1 FORMAC Symbolic Mathematics Interpreter*, IBM, Hawthorne, New York.
53. Zienkiewicz, O. C. (1971). *The Finite Element Method in Engineering Science*, McGraw-Hill, London.
54. Daly, P. and Helps, J. D. (1972). 'Direct method of obtaining capacitance from finite-element matrices', *Electronics Letters*, **8**, 5, 132–133.

Finite Elements in Electrical and Magnetic Field Problems
Edited by M. V. K. Chari and P. P. Silvester
© 1980, John Wiley & Sons Ltd.

Chapter 8

Transient Solution of the Diffusion Equation by Discrete Fourier Transformation

P. J. Lawrenson and T. J. E. Miller

8.1 INTRODUCTION

In the numerical solution of low-frequency electromagnetic fields relatively little attention has been given to transient problems. This has been due partly to the importance attributed to time-invariant problems (e.g. field strength or inductance calculations) or steady-state harmonic problems (e.g. eddy-current loss determination) and partly to lack of effective methods for solving transient problems. Even for steady-state and time-invariant problems much remains to be done in developing or, at least, improving methods. For magnetostatic problems, although various effective approaches by finite-element, finite-difference and integral methods are now available for two- and three-dimensional cases, including the effects of nonlinear material characteristics, increased efficiency in order to deal with larger and more complex cases is desirable. Steady-state, time-dependent (harmonic) linear two-dimensional problems present no great difficulty, although incorporation of nonlinearities can be troublesome. The treatment of three-dimensional cases gives rise to vector problems for which, even in the linear situation, only limited efforts and progress have been made so far. For transient problems in which time appears explicitly, time discretization (stepping) techniques, which involve a complete solution of the spatial variables over the whole region for each time step, have been explored but only to a limited extent.

The overall state of development of numerical methods and the current interest in various technological problems are such that increasing attention is being given to numerical solutions of transient problems. It is, therefore, opportune to discuss possible approaches to such problems and, in particular, to draw attention to the scope and usefulness of the numerical applications of Fourier transform methods. Fourier transform methods are standard and are very widely used in high-frequency (communication) areas of electrical engineering, but their significance has been little appreciated at lower frequencies so far. As will be shown, they are of considerable power and value in the low-frequency area. They can provide for the efficient solution of linear transient problems (for any impressed time function that may be of interest), provided only that the associated steady-state, constant-frequency, harmonic excitation problem can be solved. This latter problem can be solved by finite-element, finite-difference, integral, or analytical means (or, indeed, by experimental means). In other words, discrete Fourier transform techniques provide for the synthesis of solutions to any transient linear field problem when the steady-state eddy-current solution for the same geometry and media (i.e. for the same space-dependent parameters) is available. This technique can be used freely in all two-dimensional cases, and also in those three-dimensional cases for which solution methods exist. Boundary shapes, in principle, present no difficulties, and different field regions may be of different permeabilities and conductivities. The only limitation is that nonlinear properties of the media cannot be dealt with.

The power and flexibility of the method come from the numerical evaluation of the transform in discrete form by means of the so-called fast Fourier transform algorithm.[3] An outline of the fast Fourier transform is presented in Section 8.3 after a brief survey in Section 8.2 of other approaches to the solution of transient problems. Particular stress is laid throughout on the digital formulation of the approach and the consideration involved in the practical application of the method. Discussion of these aspects appears to have been particularly limited previously. The questions arising are illustrated in relation to two problems of practical importance. The first of these is the calculation of the transient magnetic field existing in the field winding of a superconducting a.c. generator following a step change in armature current. The second example is the calculation of the transient current in a rectangular conductor in a semi-closed slot, following the application of a voltage whose waveshape incorporates most of the features commonly found in voltage transients in electrical machines. The two examples serve to bring out several important practical aspects.

8.2   METHODS FOR TRANSIENT DIFFUSION

Transient fields in linear materials obey the diffusion equation which, in terms of magnetic field strength $\mathbf{H}$, is given by

$$\text{curl curl } \mathbf{H} = -\frac{\mu}{\rho} \frac{\partial \mathbf{H}}{\partial t}. \tag{8.1}$$

The three main methods for solving Equation (8.1) are (a) purely analytical methods, (b) time-stepping numerical methods, and (c) integral transform methods. Purely analytical solutions are possible only when the geometry of the problem is simple, and usually not more than two field components can be handled. The method is applicable in practice only to step or impulse excitation functions or to combinations of these. The solution obtained is usually a sum of exponential terms e^{-t/τ_n} (where the τ_n have to be determined from the roots of a transcendental equation) and may be slow to converge. Exact analytical solutions have been limited to linear problems,[2,8,12] though some approximate solutions to nonlinear ones may prove feasible.

Time-stepping numerical methods,[4,5] and the Alternating Direction Implicit method,[14] are of much greater scope and power. They can handle more complicated geometries, with all field components present, and can deal with general excitation functions. Also, they are the only methods which, at present, can be applied to nonlinear problems. So far, however, it appears that their application to low-frequency electromagnetic field problems has been in connection with harmonic excitation,[6,7,15] the reason for their introduction being to overcome the difficulties of handling the nonlinear iron characteristic.

Unfortunately, time-stepping methods can be very time-consuming. This is due to the large numbers of simultaneous equations which must be solved at each time step, and to the related problems of convergence and stability (though the latter is considerably eased by the use of implicit methods). Also, the entire field solution must be repeated, *ab initio*, for every different excitation function to which the response is required. Further, it is inherent that the field at all points has to be evaluated even when only a few local values are needed.

The transform method is free of most of these limitations. Its application has up to now been in purely analytical field solutions, and the commonest transforms have been those of Fourier and Laplace. The method can, however, be applied just as well to numerically formulated problems. By effectively separating the space-variation from the time-variation, it uses a 'once for all' space solution in the form of a frequency response for the field at all points of interest. From this the transient response can be obtained by inverse transformation for virtually any excitation function. The fast Fourier transform algorithm enables the transient calculation to be made quickly and efficiently.

The frequency response function of the field quantity in question is obtained from the solution of the Helmholtz equation. Both analytical and powerful numerical methods for solving this equation already exist and are being developed further. The frequency response is often also of interest in its own right and is helpful in interpreting the diffusion properties of the field. Even when the only approach to the frequency response is by direct physical measurement it may still provide an attractive overall approach to the evaluation of transient behaviour to various excitation functions, including those which cannot easily be generated under test conditions.

8.3   THE FOURIER TRANSFORM METHOD

Fourier transformation of Equation (8.1) replaces the operator $\partial/\partial t$ by $j2\pi f$ to give the complex Helmholtz equation:

$$\text{curl curl } \mathbf{H}(f) = -j2\pi f \frac{\mu}{\rho} \mathbf{H}(f) \qquad (8.2)$$

which in transient problems is to be solved for a range of values of f. The solution can be expressed as a phasor:

$$H(f) = S(f)\hat{H}_0 e^{j2\pi f t} \qquad f \sim (-\infty, \infty) \qquad (8.3)$$

where $\hat{H}_0 e^{j2\pi f t}$ is the excitation function phasor, $H(f)$ is the phasor value of any component of $\mathbf{H}(f)$ at any point, and $S(f)$ is a complex frequency-response function characterizing the field at each point. $S(f)$ depends only on the 'space' part of the problem and its particular importance is that the response $h(t)$ to *any* excitation function $h_0(t)$ can be obtained[10] readily as

$$h(t) = F^{-1}\{S(f)H_0(f)\} \qquad (8.4)$$

where $H_0(f)$ is the Fourier transform of $h(t)$ given by Equation (8.5) and the quantities in it are written, for convenience, in normalized form. For example, $h(t)$ denotes the field strength normalized to $\hat{H}_0$. All the results in the following sections will be presented in normalized units.

The inverse transformation F^{-1} is computed efficiently by means of the fast Fourier transform (FFT) algorithm. An important advantage of the method is immediately clear in that the same function $S(f)$ can be used for all excitation functions $h_0(t)$.

The method has the further advantage that the transient response at only one or a few local field points can be evaluated independently of all other field points by using the frequency response functions appropriate to the points of interest. This facility is not readily available with direct methods. Even when the field transient is required at a large number of points, requiring repeated application of Equation (8.4), the computation can be done quickly and efficiently using the FFT.

The flexibility of the transform method is further enhanced by the ease with which the excitation transform $H_0(f)$ can be obtained from $h_0(t)$ by means of the FFT. This extends the application of the method to excitation functions of any waveshape.

8.3.1   The discrete Fourier transform

The classical Fourier transform pair, given by

$$h(t) = F^{-1}\{H(f)\} = \int_{-\infty}^{\infty} H(f)e^{j2\pi f t}\,df \qquad t \sim (-\infty, \infty) \qquad (8.5)$$

$$H(f) = F\{h(t)\} = \int_{-\infty}^{\infty} h(t)e^{-j2\pi f t}\, dt \qquad f \sim (-\infty, \infty)$$

is continuous and defined on infinite domains, and for purposes of computation by digital computer they are approximated by the so-called discrete Fourier transform pair (DFT) given by

$$h_d(k) = h_d(k\Delta t) = \frac{1}{T}\sum_{l=0}^{N-1} H_d(l)e^{j2\pi kl/N} \qquad k = 0, 1, 2, \ldots, N-1$$

(8.6)

$$H_d(l) = H_d(l\Delta f) = \sum_{k=0}^{N-1} h_d(k)e^{-j2\pi kl/N} \qquad l = 0, 1, 2, \ldots, N-1$$

in which $h(t)$ and $H(f)$ are each represented by the N samples of $h_d(k)$ and $H_d(l)$ respectively. The inverse DFT of $H_d(l)$ is $h_d(k)$ which approximates $h(t)$ for $t < T/2$, where $T = N\Delta t$ is the 'length' or duration of the sampled time response and Δt is the time interval between samples. Similarly $H_d(l)$ approximates $H(f)$ for $f < F/2$ where $F = N\Delta f$ is the 'length' or bandwidth of the sampled frequency response and Δf is the spacing between samples in the frequency domain.

The functions $h_d(k)$ and $H_d(l)$ are periodic, of periods T and F respectively, and are formed by first 'aliasing' and then sampling their continuous equivalents, as shown in Figure 8.1.

First the periodic 'aliased' functions

$$h_a(t) = \sum_{m=-\infty}^{\infty} h(t+mT)$$

(8.7)

$$H_a(f) = \sum_{m=-\infty}^{\infty} H(f+mF)$$

are formed by the repetition at intervals of T and F of the central or more significant parts of $h(t)$ and $H(f)$, 'folded' about the frequency $F/2$ as in Figure

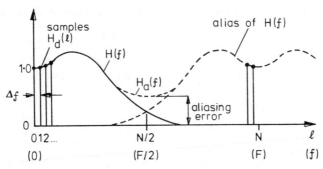

Figure 8.1 Construction of the aliased and sampled frequency-response $H_d(l)$ from $H(f)$

8.1. Then one period of each of the aliased functions is sampled at N equally spaced points:

$$h_d(k) = h_a(k\Delta t) \qquad k = 0, 1, 2, \ldots, N-1$$

$$h_d(l) = H_a(l\Delta f) \qquad l = 0, 1, 2, \ldots, N-1. \tag{8.8}$$

Since $F = N\Delta f$, $T = N\Delta t$, and $FT = N$,[1] the lengths F and T and the sampling intervals Δf and Δt cannot be chosen independently. This has important practical consequences, as will be discussed later on. $H_d(l)$ and $h_d(k)$ form the DFT pair of Equation (8.6), which are evaluated by means of the FFT algorithm.

An important saving in computing effort results from the 'causal' property of physical systems, i.e. that $h(t) = 0$ for $t < 0$. In such systems the real and imaginary parts $R(f)$ and $X(f)$ of $H(f)$ are not independent and $h(t)$ for $t > 0$ can be computed from either part by itself, using the cosine or the sine versions of the transform. A further saving results, as shown below, from the fact that the time response must be real. These savings are valuable and the basis of them will now be demonstrated. Writing $H(f) = S(f)H_0(f) = R(f) + jX(f)$ in Equation (8.4) and taking the DFT,

$$h_d(k) = \frac{1}{T} \sum_{l=0}^{N-1} [R(f) + jX(f)] e^{j2\pi kl/N} \qquad k = 0, 1, 2, \ldots, N-1$$

$$= \frac{1}{T} \sum_{l=0}^{N-1} \left[R \cos\frac{2\pi kl}{N} - X \sin\frac{2\pi kl}{N} \right] \tag{8.9}$$

$$+ j\frac{1}{T} \sum_{l=0}^{N-1} \left[X \cos\frac{2\pi kl}{N} + R \sin\frac{2\pi kl}{N} \right].$$

Since h is real, the second sum is zero and R is even in l (i.e. symmetrical about $l = N/2$) while X is odd. The economy in data preparation follows from splitting $h_d(k)$ into odd and even parts and by making use of the well-known result[10] that $h_{Ev}(k) \longleftrightarrow R(l)$; $h_{Od}(k) \longleftrightarrow X(l)$. At this stage, it is also convenient to separate out any constant or 'd.c.' component of $h_d(k)$ because its Fourier transform is a δ-function which cannot be handled by the FFT. Thus

$$h_d(k) = h_{Od}(k) + h_{Ev}(k) + h_{dc}. \tag{8.10}$$

Evaluation of h_{dc} causes no difficulty and it is determined as $\frac{1}{2}S(0)H_0(0)$ where $H_0(0)\delta(f)$ is the Fourier transform of the d.c. component in the excitation function.[10]

If the system is causal,

$$h_d(-k) = -h_{Od}(k) + h_{Ev}(k) + h_{dc} = 0 \tag{8.11}$$

so that

$$h_d(k) = 2[h_{Ev}(k) + h_{dc}] \qquad k < \frac{N}{2}(t > 0). \tag{8.12}$$

Only $h_{Ev}(k)$ needs to be evaluated and for this only $R(l)$ is required. In the summation, the evenness of $R(l)$ can be exploited to dispense with half the samples of it. Thus from Equation (8.9)

$$h_{Ev}(k) = \frac{1}{T} \sum_{l=0}^{N-1} R(l) \cos\frac{2\pi kl}{N}$$

$$= \frac{2}{T} \left\{ \frac{R(0)}{2} + \sum_{l=1}^{N/2-1} R(l)\cos\frac{2\pi kl}{N} + \frac{R(N/2)}{2}\cos \pi k \right\} \qquad (8.13)$$

in which the expansion has made use of the property $R(-l) = R(l) = R(N-l)$. The data required to evaluate Equation (8.13) is only a quarter of that needed to evaluate Equation (8.9).

In the following practical applications, the use of the cosine transform formulated in this way (Equation (8.13) is implicit, and the lengths of the frequency and time responses will be referred to as $F/2$ and $T/2$ respectively.

8.3.2 Practical procedure for applying the FFT

The first step is to form the product $H(f) = S(f)H_0(f)$. $S(f)$ will have been obtained by solving Equation (8.2) or by experiment, while $H_0(f)$ will have been determined either analytically or by FFT from the excitation function to which the response is required. The spectrum length $F/2$ is fixed at a frequency high enough so that all significant parts of $H(f)$ are included. Since $\Delta t = 1/F$, this decision is equivalent to the choice of the time sampling interval Δt. It is often quicker and more convenient to judge an appropriate value for Δt and subsequently to check that $H(f)$ has no significant values above the resulting 'folding' frequency $F/2$ (i.e. in accordance with the sampling theorem, that $F/2 > F_N$, the Nyquist frequency). The important point in this choice is that truncation in one domain causes a ripple in the other (Gibbs' phenomenon) so that if $F/2 = N\Delta f/2$ is too small, a ripple appears in $h_d(k)$. The ripple is confined to samples near the beginning and end of the period $T/2$ and it can often be tolerated in the interest of keeping N as small as possible.

The next step is to choose Δf, which is equivalent to choosing the length $T/2$ of the time response. This can be slightly more difficult than choosing Δt because often the 'settling time' of the response is not known beforehand. A graph of $H(f) = S(f)H_0(f)$ is a help in estimating the fineness with which Δf should be sampled but often this graph is not immediately available. Usually only the graph of $S(f)$ has been plotted at this stage. When dealing with smoothly varying nonperiodic time functions a reasonable initial choice of Δf is one which gives fine sampling of $S(f)$ such that the difference between adjacent samples is less than 2%. When dealing with excitation functions which have a stronger periodic content (and, therefore, a 'peaky' frequency response $H_0(f)$), a smaller Δf will be needed. It is always better to have Δf too small rather than too large. Examples of both the above types of excitation function are given in Section 8.4.

The number of samples $N/2$ required is consequent upon the choice of Δt and Δf, since $N = 1/\Delta f \Delta t$. The computing time and storage requirements depend solely on N (which for greatest computing efficiency[1] should be an integral power of 2), and thus impose an ultimate limit on the fineness with which the frequency and time functions can be sampled.

8.4   APPLICATIONS

8.4.1   Screening the field winding of a superconducting a.c. generator

As a first example of the application of the FFT method, the transient magnetic field at the rotor winding of a projected $2p$-pole superconducting a.c. generator will be calculated. In this generator, the superconducting winding on the rotor is surrounded by two concentric cylindrical eddy-current screens whose purpose is to prevent sudden or rapid changes of magnetic field near the superconductor, excited by transient changes in armature current or rotor speed. A severe case of such transients is a step change of armature current caused, for example, by a fault or a switching operation.

The calculation of the transient field at the superconductor involves the solution of equation (8.1) in the idealized model of the machine shown in Figure 8.2, in which a two-dimensional formulation is chosen because of its relative simplicity and because useful general properties of the screening system can be derived from it and interpreted without much difficulty. The first step is to determine the field under steady-state conditions with excitation by a synchronously rotating, harmonically varying current sheet $K \sin p\theta \, e^{j2\pi f t}$. The

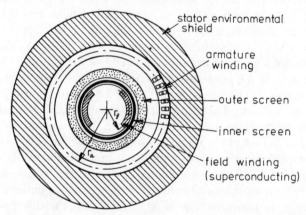

Figure 8.2   Cross-section of superconducting alternator showing windings and screens

basic field equation (8.2) in cylindrical coordinates is

$$\left[\frac{d^2}{dr^2}+\frac{1}{r}\frac{d}{dr}-\left(\frac{p^2}{r^2}+j2\pi f\frac{\mu}{\rho}\right)\right]H=0. \tag{8.14}$$

Attention is concentrated on the field component H_θ at the superconducting winding $(r=r_f)$ and the solution for this can be written as[9]

$$H_\theta=-\frac{K}{2}\left(\frac{r_f}{r_a}\right)^{p-1}S(f)e^{j2\pi ft}\sin p\theta \tag{8.15}$$

where $S(f)$, the screening ratio frequency response function (appearing also in Equation (8.4)), is the ratio between the phasor value of H_θ at frequency f and at zero frequency. It is a complex expression involving Bessel functions and characterizes the screening properties of the double screen system of Figure 8.2. The value of $S(f)$ is plotted in Figure 8.3 in the range $f=0.1$ Hz to $f\simeq$ 10 Hz. As expected with an eddy-current screen, $S(f)$ falls (i.e. the screening improves) as the frequency increases.

Two particular frequency responses of interest are shown in Figure 8.3. The first corresponds to the case where the field winding is open-circuited and has no large-scale effect on the transient fields: in this case $S(f)$ is given by the lower curve in the figure. In the second case the field circuit is closed (as would normally be the case) through a resistance such that its time constant L_f/R_f is much longer than the time taken for the field to diffuse through the screen. This situation is represented by introducing a diamagnetic boundary ($B_r=0$ at $r=r_f$) to represent the property that, in the direct axis, the transient flux cannot link the field winding during a short period of time compared with L_f/R_f. The resulting values of $S(f)$ are larger than for the case with the field winding open-circuited. It is apparent that the effectiveness of the screening system is degraded by closing the field winding.

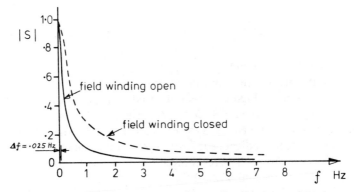

Figure 8.3 Magnitude of screening-ratio frequency-response function

While the examination of $S(f)$ is instructive and reveals several general properties of the screening system and the degradation of screening due to a closed field winding, its main importance here is the frequency response function from which transient responses can be calculated using Equation (8.4). In the particular case of the step change of armature current, the excitation function can be written as

$$H_{\theta 0}(t) = -\frac{K}{2}\left(\frac{r_f}{r_a}\right)^{p-1} \sin p\theta \ u(t) = \hat{H}_{\theta 0} u(t) \tag{8.16}$$

where $u(t)$ is the unit step function. The necessary Fourier transform of this can conveniently be obtained analytically in this case as[10]

$$H_{\theta 0}(f) = -\frac{K}{2}\left(\frac{r_f}{r_a}\right)^{p-1} \sin p\theta \left[\pi\delta(f) + \frac{1}{j2\pi f}\right] \tag{8.17}$$

and Equation (8.4) immediately gives the transient response as

$$H_{\theta}(t) = -\frac{K}{2}\left(\frac{r_f}{r_a}\right)^{p-1} \sin p\theta F^{-1}\left\{\left[\pi\delta(f) + \frac{1}{j2\pi f}\right]S(f)\right\}$$

$$= h_{\theta}(t)\hat{H}_{\theta 0} \tag{8.18}$$

where $h_{\theta}(t)$ is the response normalized to $\hat{H}_{\theta 0}$. The transient response contains a 'd.c.' term which is separated immediately:

$$h_{dc} = F^{-1}\{\pi\delta(f)S(f)\} = \tfrac{1}{2}S(0) = \tfrac{1}{2} \tag{8.19}$$

while the remainder of the inverse transformation is computed by means of the FFT as $h_d(k)$, i.e.

$$F^{-1}\left\{\frac{S(l\Delta f)}{j2\pi l\Delta f}\right\} = \frac{4}{T}\left\{\frac{R(0)}{2} + \sum_{l=1}^{N/2-1} R(l)\ \cos\frac{2\pi kl}{N}\right.$$

$$\left. + \frac{R(N/2)}{2}\cos \pi k\right\} \qquad k = 0, 1, 2, \dots, N-1 \tag{8.20}$$

where $R(l) = \mathrm{Re}[S(l\Delta f)/j2\pi l\Delta f]$.

Following the procedure outlined in Section 8.3, the first step is to select $F/2$ (or Δt). Inspection of Figure 8.3 shows that $[S(f)]$ is less than 5% of its maximum magnitude when $f > 6.5$ Hz and since $1/j2\pi f$ is also a decreasing function of f, this is taken as a 'safe' value for $F/2$. The product $H(f) = S(f)H_0(f)$ has a steep gradient near $f = 0$, necessitating fine sampling, and a value of $\Delta f = 0.025$ Hz is chosen. The resulting value of $N/2$ is $F/2\Delta f = 6.5/0.025 \cong 256$; giving $\Delta t = 1/F = 77$ ms and $T/2 = 20$ s. The resulting inverse transform $h_d(k)$ is shown in Figure 8.4 for both cases (field winding open and closed) and it can be seen that both the final d.c. error and the 'initial sample error' are negligible, indicating a

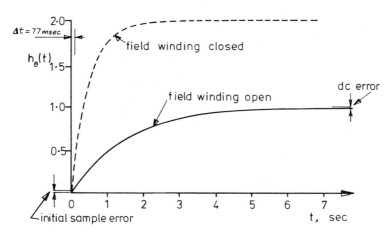

Figure 8.4 Transient response of tangential field strength near superconducting winding, following step change in armature current

satisfactory choice of Δf and Δt respectively. (The actual values of d.c. error and the initial sample error are, respectively, 0.01% and 0.37% for the open-circuit field winding and 0 and 0.33% for the closed field winding. With Δf reduced to 0.01 Hz, the corresponding values of these errors are 0 and 1.42% for the open field winding and 0.01% and 3.9% for the closed field winding. The increased initial sample error with $\Delta f = 0.01$ Hz reflects the time-domain ripple caused by truncating $H(f)$ at too low a frequency.)

The values of $h_{\theta d}(k) + h_{dc}$ are normalized to $\hat{H}_{\theta 0}$, the maximum d.c. value of the field when the field winding is open. Figure 8.4 shows the doubling effect caused by the diversion of direct-axis flux round the field winding. The initial slopes of the curves show a much increased rate of rise of field when the field winding is closed, and this illustrates the degradation of screening (in the direct axis) caused by the closed field winding.

8.4.2 Rectangular conductor in a semi-closed slot

As a second example of the application of the FFT to a transient diffusion problem, the current in a rectangular conductor in a semi-closed slot will be calculated for an applied voltage waveform which provides an example of a slightly more complicated excitation function than that of the previous section.

The configuration to be studied is shown in Figure 8.5. A copper conductor of rectangular section fills a semi-closed slot in an infinitely permeable material. The frequency-response function most appropriate for characterizing this field situation is the complex admittance $Y(f)$ of the conductor measured between its ends. Once this is known, the current $i(t)$ following the application

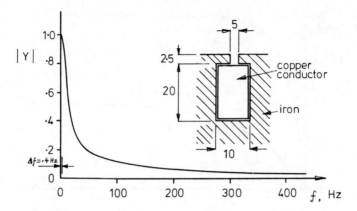

Figure 8.5 Semi-closed slot geometry and magnitude of admittance as a function of frequency. (Dimensions are in mm)

of any nonperiodic voltage $v_0(t)$ is given by Equation (8.4) in the form

$$i(t) = F^{-1}\{Y(f)V_0(f)\} = F^{-1}\{I(f)\} \tag{8.21}$$

where $V_0(f)$ is the Fourier transform of $v_0(t)$. The magnetic field at any point can be calculated from $i(t)$.[13]

In this example the admittance function $Y(f)$ can be computed by means of a finite-element or finite-difference formulation or from an approximate analytical field solution.[13,11] Figure 8.5 shows the magnitude $Y(f)$ normalized to the d.c. conductance for the conductor whose dimensions are also given there, as calculated by Rolicz's method.[11]

Transient response

The current $i(t)$ will be determined for the particular voltage excitation

$$v_0(t) = [1 + e^{-at} \cos \omega_0 t] u(t) \tag{8.22}$$

i.e. a combination of a 'd.c.' step and a cosine wave which decays to about 0.01 in 10 cycles ($a = 23.04$; $\omega_0 = 314$ rad/s); see Figure 8.6(a). This excitation function possesses most of the essential features to be found in voltage transients in electrical machines, yet its Fourier transform is simple enough to be calculated directly from Equation (8.5).[10] Thus

$$V_0(f) = \pi\delta(f) + \frac{1}{j\omega} + \frac{a + j\omega}{(a + j\omega)^2 + \omega_0^2} \tag{8.23}$$

with $f = \omega/2\pi$. The first two terms are the FT of the step, while the third is the FT of the decaying cosine wave.

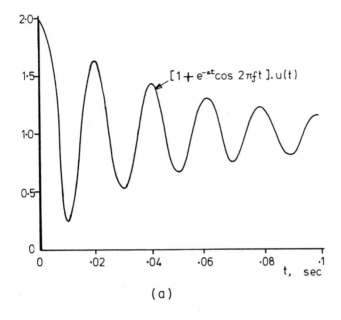

$$[1 + e^{-\alpha t}\cos 2\pi f t].u(t)$$

(a)

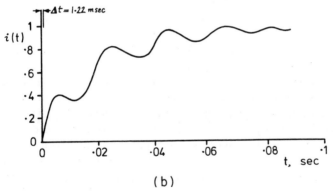

(b)

Figure 8.6 (a) Voltage excitation (forcing) function. (b) Transient response of current in slot

The transient current $i(t)$ is now computed using Equation (8.21). It will be clear that in general the response to the d.c. step will not require the same number of samples in the time domain as the rapidly varying response to the cosine wave, of which 10 full cycles are to be represented. For this reason the responses to the two parts of $v_0(t)$ are evaluated separately as $i_u(t)$ and $i_c(t)$ and the total response is taken as the sum $i_u(t) + i_c(t)$. The linearity of the transform and of the field solution makes this process valid. The process of obtaining the step response is similar to that in the previous section and will not be discussed beyond stating the values $\Delta f = 0.4$ Hz, $N/2 = 1024$, $\Delta t = 1.22$ ms. $T/2 = 1.25$ s, $F/2 = 410$ Hz.

To obtain the response $i_c(t)$ to the decaying cosine wave the first step is to decide the folding frequency $F/2$ such that the frequency response function $I_c(f) = Y(f) V_0(f)$ has no significant components at frequencies above it. It is more convenient in this example, as mentioned in Section 8.3, to choose Δt rather than F (or $F/2$), and taking 16 samples per cycle of the 50 Hz component in the excitation function gives $\Delta t = 1.25$ ms, $F/2 = 400$ Hz. It can be checked that at and above this frequency, the function $I_c(f)$ has negligible amplitude, so that the truncation can be done safely. The next step is to choose Δf (or $T = 1/\Delta f$) and without plotting the graph of $I_c(f)$ a conservative value must be tried first. To give fine sampling of $Y(f)$, a value of $\Delta f = 0.4$ Hz is initially chosen. The number of samples required is now given as $N/2 = 1/2\Delta f \Delta t = 1000$, or 1024 as an integral power of 2, giving, finally, $\Delta t = 1.22$ ms.

The sum of the resulting response $i_c(t)$ and the step response $i_u(t)$ is shown in Figure 8.6. Two main errors are present in this result. The initial sample error is 4%, of which 3.2% is contributed by $i_c(t)$. This error, which is due to corruption of the frequency response caused by having $F/2$ too small, could be reduced by increasing $F/2$ (or reducing Δt); for example, with $\Delta f = 1.0$ Hz and $\Delta t \cong 0.488$ ms the initial sample error in $i_c(t)$ falls to only 1.3%. The second error is the d.c. error of about 2% which is contributed almost entirely by the step response. Again it is due to corruption of the frequency response but in this case because Δf is too large. (Equivalently, $T/2$ is too small and the error is the time-domain aliasing error in the sense illustrated for $H(f)$ in Figure 8.1.) By reducing Δf to 0.1 Hz, the d.c. error is virtually eliminated.

The elimination of the errors by separate and different adjustments of Δf in the computation of $i_c(t)$ and $i_u(t)$ takes advantage of the division of the total response into these two components, to limit N and therefore the computing time. In many cases this division is not possible and the simultaneous reduction of Δf and Δt to improve accuracy must be paid for by an increase in N and the computing time.

8.5   CONCLUSIONS

The Fast Fourier Transform can be used with advantage to solve transient electromagnetic diffusion fields. Particular attention must be paid to the various considerations arising in the practical applications of the method as described in the preceding sections.

The effectiveness of the method arises basically because the total solution can be broken down into two separate stages, one concerned with space considerations and the other with time, each stage being much easier to handle than the total situation. In more detail, the approach is effective since

(i) The time-stepping part of the solution, which gives rise to stability problems, is eliminated.

(ii) The complete solution for the time-dependent part of the problem (i.e. the complete transient response) can be accomplished quickly and efficiently. The computing time for this stage is likely to be only a small fraction of that for the space-dependent part of the solution. Indeed, dedicated processors exist for carrying out just this evaluation.

(iii) The space dependence of the problem is generally the most troublesome and time consuming but, even for this stage of the solution, advantage can be taken of various established and effective methods.

The major economy with the FFT method is achieved when a particular field configuration has to be studied subject to various excitation functions. Since only a *single* frequency-response function $S(f)$ needs to be evaluated, this is used repeatedly with the FFT for each time function. However, significant economies should be achievable even when only a single time function is in question (the method being applied exactly as described). This will be the case particularly when the excitation functions have complicated wave shapes whose Fourier transforms are not immediately available or cannot be worked out analytically.

Further economies in computing time can be achieved by taking advantage of the generally simple form of $S(f)$, and employing some suitable algorithm to determine an adequate representation of $S(f)$ from the minimum number of increments in f. (For subsequent application of the FFT, the sampling can be as fine as is required.)

The general smoothness of $S(f)$ often encountered in diffusion problems enhances the computational efficiency of the method. Particularly when the excitation function transform is also a smooth function, as is often the case with, for example, suddenly applied d.c. excitation functions, the sampling interval in the frequency domain can be relatively coarse, making it possible to achieve fine sampling in the time domain without requiring excessive extra computing effort.

REFERENCES

1. Bergland, G. D. (1969). 'A guided tour of the fast Fourier transform', *IEEE Spectrum*, **6**, 41–52.
2. Carslaw, H. S. and Jaeger, J. C. (1959). *Conduction of heat in solids*, 2nd Edition, Clarendon Press, Oxford.
3. Cooley, J. and Tukey, J. (1965). 'An algorithm for the machine calculation of complex Fourier series', *Math. Comp.*, **19**, 297–301.
4. Crank, J. and Nicolson, P. (1947). 'A practical method for numerical inversion of solutions of partial differential equations of the heat-conduction type', *Proc. Camb. Soc.*, **43**, 50–67.

5. Dufort, E. C. and Frankel, S. P. (1953). 'Stability conditions in the numerical treatment of parabolic differential equations'. in *Mathematical tables and aids to computation*, Vol. 7, US NAC–NRC, Washington, USA, 135–152.

6. Foggia, A., Sabonnadière, J. C., and Silvester, P. (1975). 'Finite-element solution of saturated travelling magnetic field problems', *IEEE Transactions*. Paper T75 030-2, PES Winter Meeting, New York, 1975.

7. Gillott, D. H. and Calvert, J. F. (1965). 'Eddy-current loss in saturated solid magnetic plates, rods and conductors', *IEEE Trans.*, **MAG-1**, 126–137.

8. Miller, K. W. (1947). 'Diffusion of electric current into rods, tubes, and flat surfaces', *Trans. Amer. Inst. Elec. Engrs.*, **66**, 1496–1502.

9. Miller, T. J. E. and Lawrenson, P. J. (1976). 'Penetration of transient magnetic fields through conducting cylindrical structures, with particular reference to super-conducting ac machines', *Proc. IEE*, **123**, No. 5, 437–443.

10. Papoulis, A. (1962). *The Fourier integral and its applications*, McGraw-Hill, New York.

11. Rolicz, P. (1976). 'Transient skin-effect in a rectangular conductor placed in a semi-closed slot', *Archiv für Elektrotechnik*, **57**, 329–338.

12. Smythe, W. S. (1950). *Static and dynamic electricity*, 2nd Edition, McGraw-Hill, New York.

13. Swann, S. A. and Salmon, J. W. (1962). 'Effective resistance and reactance of a solid cylindrical conductor placed in a semi-closed slot', *Proc. IEE*, **109**, 611–615.

14. Peaceman, D. W. and Rachford, H. H. (1955). The numerical solution of parabolic and elliptic differential equations. *J. Soc. Ind. Appl. Math.*, **3**, 28–41.

15. Lim, K. K. and Hammond, P. (1972). Numerical method for determining the electromagnetic field in saturated steel plates, *Proc. IEE*, **119**, 1667–74.

Finite Elements in Electrical and Magnetic Field Problems
Edited by M. V. K. Chari and P. P. Silvester
© 1980, John Wiley & Sons Ltd.

Chapter 9

Mutually Constrained Partial Differential and Integral Equation Field Formulations

B. H. McDonald and A. Wexler*

9.1 INTRODUCTION

In the mathematical solution of engineering continuum problems two distinct schools exist. There are those who attempt solution of the partial differential equations directly, and there are others whose preference it is to make use of an elemental solution (known as the Green's function) and to seek the field directly by integration of a known source distribution or, if only boundary conditions are stated, to seek the field indirectly by first solving an integral equation for the source distribution needed to sustain the stated boundary conditions. The former method is sometimes known as the field approach and the latter as a source distribution technique. We shall refer to these methods as the partial differential equation (PDE) and integral equation (IE) methods respectively.

Hammond,[1] with specific reference to electromagnetism, has placed these dual approaches in a historical perspective:

> '... The history of electromagnetic investigation is the history of the interplay of two fundamentally different modes of thought. The first of these, the method of electromagnetic fields which ascribes the action to a continuum, is associated with such thinkers as Gilbert, Faraday and Maxwell. The second, the method of electromagnetic sources, concentrates attention on the forces between electric and magnetic bodies and is associated with Franklin, Cavendish and Ampère ... field problems are conveniently handled by differential equations and sources by integral equations ... the rigid distinction between field and source methods can be discarded only when it is seen that there is a freedom of choice...'

* The research reported in this chapter was supported by the National Research Council of Canada.

Many analysts tend to settle upon a choice of one of the methods with a commitment that virtually excludes the other from consideration. This is regrettable because, as one might expect, each approach has its own virtues and shortcomings and the selection should be based upon the individual requirements of the problem at hand. Moreover, not only might one expect one or the other approach to be preferred for a particular problem but also it is possible that a given problem might be most expeditiously addressed by using a partial differential equation (PDE) approach over one part of the region and an integral equation (IE) approach over the remainder.

This chapter has the following goals:

(i) to indicate where a mix of IE and PDE formulations can be advantageous;
(ii) to describe algorithms that permit a mixed approach to be conveniently accomplished through mutually constraining the differential and integral operators in various ways;
(iii) to encourage the generalization of the finite-element concept to include its application to the solution of IEs alone and in hybrid situations; and
(iv) to indicate the essential unity of a number of apparently disparate approaches described by various authors.

The authors,[2] and with others,[3,4,5] have previously reported on various aspects of these goals. Salient features of these papers will be briefly summarized in order to clarify the differences between the methods and their relative advantages. The mutual constraint of PDE and IE formulations, both defined in variational terms, has not—to our knowledge—been previously reported and is therefore described in some detail herein.

This chapter continues with brief reviews of the basic PDE and IE formulations, Rayleigh–Ritz discretization, functionals for nonself-adjoint operators and Galerkin equivalency, and moment methods. This is followed by a description of the Lagrange constraint method in finite-element systems, the PDE formulation using a nonvariational (i.e. moment) constraint, the variational procedure for the finite-element solution of IEs, and the mutual constraint of PDE and IE variational forms. We conclude with a number of examples that indicate the utility of the hybrid PDE–IE approach, the importance of using continuous trial functions rather than the usual pulse functions in IE solution, and the extremely rapid convergence obtained when the asymptotic form of the edge singularity is included in the trial function set when appropriate. Reference is made to a numerical experiment that displays the superior convergence of the variational approach compared to a moment method using delta function testing.

9.1.1 Basic Equations

As specific examples, this section concentrates on the scalar Poisson

$$-\nabla \cdot (\epsilon \nabla \phi) = \rho \tag{9.1}$$

and Helmholtz

$$-\nabla^2\phi - k^2\phi = \frac{\rho}{\epsilon} \tag{9.2}$$

partial differential equations in a region of space R where ϵ is the dielectric constant (or magnetic permeability, etc.) which may be a tensor (in which case we have an anisotropic medium), k is the propagation constant, ρ is a given source distribution in R, and ϕ is the scalar potential to be determined.

What follows is not restricted to scalar problems. The theory applies to vector situations as well as to other PDEs, e.g. to those equations that govern linear elasticity, magnetic vector potential, etc.

Equations (9.1) and (9.2) must satisfy certain conditions on S, the boundary of R. The Dirichlet condition is

$$\phi(\mathbf{r}) = g(\mathbf{r}) \qquad \mathbf{r} \text{ on } S_1 \tag{9.3}$$

the Neumann condition is

$$\hat{\mathbf{n}} \cdot (\epsilon(\mathbf{r})\nabla\phi(\mathbf{r})) = h(\mathbf{r}) \qquad \mathbf{r} \text{ on } S_2 \tag{9.4}$$

and the mixed condition is

$$\hat{\mathbf{n}} \cdot (\epsilon(\mathbf{r})\nabla\phi(\mathbf{r})) + \alpha(\mathbf{r})\phi(\mathbf{r}) = h(\mathbf{r}) \qquad \mathbf{r} \text{ on } S_3 \tag{9.5}$$

of which the Neumann condition is but a special case.

The unit vector $\hat{\mathbf{n}}$ is taken to point out of R and the boundary $S = S_1 \cup S_2 \cup S_3$ is shown in Figure 9.1. Part or all of S may be at infinity, in which case R is an unbounded region.

Equations (9.1) and (9.2) have solutions in free-space that are given by

$$\phi(\mathbf{r}) = \frac{1}{\epsilon_0} \int_R \rho(\mathbf{r}') \, G(\mathbf{r}|\mathbf{r}') \, d\Omega' \tag{9.6}$$

where the Green's function in, say, three-dimensional space is given by

$$G(\mathbf{r}|\mathbf{r}') = \frac{\exp(-jk|\mathbf{r}-\mathbf{r}'|)}{4\pi|\mathbf{r}-\mathbf{r}'|} \tag{9.7}$$

and where $\mathbf{r}$ is the vector position of the point at which potential $\phi(\mathbf{r})$ is to be found, $\mathbf{r}'$ is the vector position of the source, and $|\mathbf{r}-\mathbf{r}'|$ is the distance between the two points. The case $k=0$ corresponds to the Green's function for the Poisson equation. Note that the Green's function is singular when $\mathbf{r}=\mathbf{r}'$.

The integration of (9.6) produces a potential which is continuous and finite, even at points within regions containing sources. The Green's function singularity is integrable, but care must be taken when performing integrations to ensure accurate results.

As a particular example, it is known that there is a unique source distribution required to sustain a specified Dirichlet potential on a surface. We may

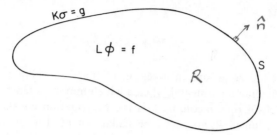

Figure 9.1 Coupled PDE and IE operator equations

seek this source distribution as a solution to the given problem, for once we know the source distribution we may compute the potential at any point in space by performing an integration (9.6). The analyst can think of placing an estimate to this unknown source distribution (e.g. electric charge, etc.) upon the contour at which the Dirichlet potential is required and then adjusting it until the potential generated satisfies the requisite condition. Consider the problem in free space and let σ be the simple source placed on the contour S. If $\phi(s) = g(s)$ is the required Dirichlet condition, the Fredholm integral equation (IE) that we must solve is

$$\frac{1}{\epsilon_0} \int_S \sigma(\mathbf{r}')\ G(\mathbf{r}|\mathbf{r}')\ \mathrm{d}s' = g(\mathbf{r}) - \frac{1}{\epsilon_0} \int_R \rho(\mathbf{r}'')\ G(\mathbf{r}|\mathbf{r}'')\ \mathrm{d}\Omega'' \qquad \mathbf{r} \text{ on } S \qquad (9.8)$$

in which the right-hand side is a known quantity. This equation has a unique solution if S is a piecewise continuous contour. When σ is found the potential it generates will be continuous, and will satisfy (9.2) exactly at any point in space as well as satisfying the requisite Dirichlet condition. In the special case where there is no given source distribution in space (i.e. $\rho = 0$), Equation (9.8) reduces to the rather more familiar IE

$$\frac{1}{\epsilon_0} \int_S \sigma(\mathbf{r}')\ G(\mathbf{r}|\mathbf{r}')\ \mathrm{d}s' = g(\mathbf{r}) \qquad \mathbf{r} \text{ on } S. \qquad (9.9)$$

One of the major advantages of the source distribution technique is that we need to seek a solution for the source only over a surface in space and we do not need to discover a potential throughout the space. This reduction in dimensionality results in the manipulation of fewer unknowns in obtaining the approximate solution to a field problem. With the design of efficient algorithms, this could reduce computing costs and, perhaps even more importantly, reduce data preparation times by an order of magnitude. On the other hand, one must often deal explicitly with the Green's function singularity when solving the integral equation. Fortunately, it turns out that these singularities do not result in the 'algorithmic nightmares' suggested by Acton.[6]

It should be observed that use of the simple free-space Green's function restricts us to homogeneous, free-space regions. We cannot place an object in the region without clothing it in a source distribution that represents the discontinuity in properties of the media. More generally, if a region is continuously inhomogeneous, then one is obliged to postulate a continuous distribution of unknown sources over the region. In doing this, the IE approach no longer produces a reduction in dimensionality and so one of its principal advantages is lost.

The PDE approach can deal effectively with anisotropic and inhomogeneous media[18,25] but it is inefficient or impractical for large or infinite regions as the number of unknowns required in the numerical solution becomes too large to handle. It is clear, therefore, that an inhomogeneity in free-space might best be handled by using a PDE formulation within the inhomogeneous region and the IE formulation for the rest of space. In fact, in such problems, it is often advantageous to use these two fundamentally different approaches together in seeking the solution. The general mathematical conception follows.

Equation (9.1) or (9.2), or whichever other PDE is being considered, may be expressed as

$$L\phi = \rho \tag{9.10}$$

where L is the partial differential operator and ϕ the potential sought (a scalar or a vector), given a source distribution $\rho(\mathbf{r})$. Analogously, Equation (9.8) or (9.9) may be written in the form

$$K\sigma = g \tag{9.11}$$

where K is an integral operator and source $\sigma(s)$ is the source to be determined given the potential distribution $g(s) = \phi(s)$.

The finite-element solution of Equation (9.10) is well known and widely reported. Equation (9.11) may be solved in an analogous fashion by extension of the finite element approach to integral equations.[21,22] In such a formulation, $\sigma(s)$ need not be a real source but can be viewed as an equivalent surface distribution of the source within a region enclosed by an arbitrary surface, as seen from the outside.

A physical representation of such a situation is depicted in Figure 9.1, in which the PDE governs the field within the enclosed region and the IE caters for the field in the open region. The IE can, incidentally, also be used in closed regions[3]—particularly as a device to economize on storage requirements. From a simplistic point of view, one can consider the whole problem to consist of two equations (9.10) and (9.11) in two unknowns $\phi(\mathbf{r})$ and $\sigma(s)$. In practice, we choose to express the equivalent source distribution in terms of $\phi(s)$ and its normal derivative, corresponding to single and dipole source layers. Later in this chapter, algorithms for solving this problem are described.

9.1.2   Variational solution methods

If an operator L is self adjoint, i.e.

$$\langle Lu, v \rangle = \langle u, Lv \rangle \qquad (9.12)$$

and positive-definite

$$\langle Lu, u \rangle \begin{cases} >0 & u \neq 0 \\ =0 & u=0 \end{cases} \qquad (9.13)$$

then the function u that minimizes the functional[7,8]

$$F(u) = \langle Lu, u \rangle - \langle u, f \rangle - \langle f, u \rangle \qquad (9.14)$$

satisfies the equation

$$Lu = f \qquad (9.15)$$

in the mean. The inner product, denoted by the angular brackets, is defined by

$$\langle u, v \rangle = \int_R uv^* \, w \, d\Omega \qquad (9.16)$$

where R is the region of interest and v^* is the complex conjugate of v. The weighting function w is taken to be unity (i.e. $w = 1$) unless otherwise specified.

Suppose that u is to be approximated by a linear combination of N real-valued functions v_i with complex coefficients:

$$u(\mathbf{r}) \cong u_N(\mathbf{r}) = \sum_{i=1}^{N} c_i v_i(\mathbf{r}) = \mathbf{c}^T \mathbf{v} = \mathbf{v}^T \mathbf{c} \qquad (9.17)$$

We wish to adjust the coefficients so that u_N is a reasonable approximation to u, and we accomplish this by using (9.14) with u replaced by u_N:

$$F(u_N) = \int_R \{ (L\mathbf{v}^T \mathbf{c})(\mathbf{v}^T \mathbf{c}^*) - (\mathbf{v}^T \mathbf{c}) f^* - (\mathbf{v}^T \mathbf{c}^*) f \} \, d\Omega \qquad (9.18)$$

or, equivalently,

$$F(u_N) = \mathbf{c}^{*T} \int_R \mathbf{v} L\mathbf{v}^T d\Omega \, \mathbf{c} - \mathbf{c}^T \int_R \mathbf{v} f^* \, d\Omega - \mathbf{c}^{*T} \int_R \mathbf{v} f \, d\Omega \qquad (9.19)$$

Taking the partial derivative with respect to each coefficient and setting each to zero results in the Rayleigh–Ritz system of equations:

$$[S]\mathbf{c} = \mathbf{b} \qquad (9.20)$$

where the matrix $[S]$ is given by

$$[S] = \int_R \mathbf{v} L\mathbf{v}^T d\Omega \qquad (9.21a)$$

or

$$s_{ij} = \int_R v_i L v_j \, d\Omega \qquad (9.21b)$$

and where

$$\mathbf{b} = \int_R v f \, d\Omega \qquad (9.22a)$$

or

$$b_i = \int_R v_i f \, d\Omega. \qquad (9.22b)$$

As the operator L is taken to be self-adjoint, the matrix $[S]$ is symmetric for real L.

9.1.3 Functionals for nonself-adjoint operators

When the operator is nonself-adjoint the preceding discussion does not apply. For example, the integral operator (9.11) using (9.7) produces the following inner products:

$$\langle Ku, v \rangle = \frac{1}{4\pi} \int_R \int_R v^*(\mathbf{r}) \, u(\mathbf{r}') \frac{\exp(-jk|\mathbf{r} - \mathbf{r}'|)}{|\mathbf{r} - \mathbf{r}'|} d\Omega' \, d\Omega \qquad (9.23a)$$

and

$$\langle u, Kv \rangle = \frac{1}{4\pi} \int_R \int_R v^*(\mathbf{r}) \, u(\mathbf{r}') \frac{\exp(jk|\mathbf{r} - \mathbf{r}'|)}{|\mathbf{r} - \mathbf{r}'|} d\Omega' \, d\Omega. \qquad (9.23b)$$

Clearly, the complex time-harmonic integral operator is not self-adjoint.

However, we may use the definition of the adjoint operator[31]

$$\langle Lu, v_a \rangle = \langle u, L_a v_a \rangle \qquad (9.24)$$

to tackle this difficulty. The adjoint operator may be presented as well in terms of the adjoint problem

$$L_a u_a = f_a. \qquad (9.25)$$

Physically f_a is normally not known, and without loss of generality, we may let $f_a = f^*$.

Mathematically, we can make use of the adjoint problem (9.25) to construct a functional for nonself-adjoint operators. Consider the functional

$$F(\mathbf{u}) = \langle \mathscr{L} \mathbf{u}, \mathbf{u} \rangle - \langle \mathbf{u}, f \rangle - \langle f, \mathbf{u} \rangle \qquad (9.26)$$

where

$$\mathcal{L} = \begin{bmatrix} L & 0 \\ 0 & L_a \end{bmatrix} \tag{9.27}$$

is a matrix operator, and

$$\mathbf{u} = \begin{bmatrix} u \\ u_a \end{bmatrix} \tag{9.28}$$

and

$$f = \begin{bmatrix} f \\ f_a \end{bmatrix} \tag{9.29}$$

are column matrices of functions. A weighting function is used in the definition of the bilinear mapping of (9.26). Here, W is a constant matrix, and an appropriate form is

$$W = \begin{bmatrix} 0 & 0 \\ 1 & 0 \end{bmatrix}. \tag{9.30}$$

In proving stationarity of (9.26), one requires that, for an arbitrary $\mathbf{v}$, if $\langle \mathbf{u}, \mathbf{v} \rangle + \langle \mathbf{v}, \mathbf{u} \rangle = 0$ then $\mathbf{u} \equiv 0$. The definition given by (9.30) satisfies this condition. It may be shown[19] that equations (9.20)–(9.22), which are identical to those obtained from Galerkin's method, result.

Another example of a nonself-adjoint integral operator, which is nonself-adjoint even with $k = 0$, is that governing the field at an interface between media of differing properties. One method of handling such an interface problem was given in a previous paper.[4] A more efficient approach, using the adjoint operator, is seen to produce equations corresponding to the Galerkin method.[19] In practice, therefore, we construct the matrix equation (9.20) directly for integral equations using the Galerkin method, with the understanding that it may also be obtained at the stationary point of a functional.

Mikhlin[7] states that an operator need be neither self-adjoint nor positive-bounded-below to ensure convergence. If the operator possesses a component with such properties, and given certain uniqueness and completeness conditions, then a convergence guarantee holds. From the experience of the authors, solutions for such problems seem to converge about as well as those involving self-adjoint, positive-definite operators.

9.1.4 The moment method

The moment method[23,24] employs a set of expansion functions as in (9.17). However, the required inner products are performed with another set of functions called testing functions:

$$\langle \boldsymbol{W}, L\boldsymbol{v}^T \rangle \boldsymbol{c} = \langle \boldsymbol{W}, f \rangle. \tag{9.31}$$

The system matrix and the right-hand side forcing vector are therefore given by

$$S = \left[\int w_i L v_j \, d\Omega \right] \tag{9.32a}$$

which is generally asymmetric, and

$$\boldsymbol{b} = \left[\int w_i f \, d\Omega \right]. \tag{9.32b}$$

Matrix equation (9.31) can be solved for $\boldsymbol{c}$ and hence produces an approximation to the field. The advantage of this approach is that the testing functions can be delta functions thus reducing the inner product to a simple and inexpensive evaluation over a set of points. This can result in a significant economy in the case of the integral equation solution, in which a double surface integration is essentially reduced to a single surface integration. The disadvantages are that, even for positive-definite operators, general convergence guarantees are lacking and the method appears to be less accurate than the variational approach requiring, therefore, a greater number of equations than would the corresponding variational scheme.[21]

9.2 FINITE ELEMENTS AND MIXED FORMULATIONS

In this section, beginning with the familiar PDE finite element technique with constraints between equations, the method is extended to cater for large and open regions using boundary constraints generated by application of the Green's function (or source distribution) method. This is followed by a description of a finite-element approach for the solution of integral equations. Finally, the algorithm for the solution of partial differential equations and integral equations, mutually constrained in a variational fashion, is described.

9.2.1 Finite elements for the discretized PDE problem

The region of space over which ϕ, the solution to the PDE problem (Equations (9.1) and (9.2)), is to be obtained, is divided into M finite elements. Within each finite element, ϕ is expressed as a real, linear conbination of potentials at N node points within the element. It is convenient to use the isoparametric finite element technique[9] to accomplish this. If we let $\boldsymbol{\phi}_i$ be the vector of node potentials in element i, and $\boldsymbol{a}_i(\mathbf{r})$ be the vector of interpolatory functions for element i, we have as the approximation

$$\phi(\mathbf{r}) = \boldsymbol{a}_i^T \boldsymbol{\phi}_i \qquad \mathbf{r} \text{ in element } i. \tag{9.33}$$

For this finite element it is possible to construct the Rayleigh-Ritz matrix equation in the form, from (9.20),

$$[S_i]\phi_i = b_i. \tag{9.34}$$

By extension, each i can refer not only to a single finite element but to an aggregate of elements treated as one. Assuming that specified boundary conditions have been incorporated into the matrix equations, only the continuity of the potential and the flux at inter-element boundaries need to be enforced. Solutions can then be obtained for all node potentials. Usually, a Green's identity is applied and the resulting element boundary integrations are disregarded. This causes the interface flux continuity condition, even for fairly general media, to become a natural condition[25] and results in considerable algorithmic simplification.

Continuity is normally enforced by assigning a single potential to nodes which occupy the same physical location. This is effected by giving common nodes the same node numbers and treating them as the same node in the algorithm that assembles the system matrices. There is yet another way of looking at the situation which helps to set the stage for the mutual constraint of IE and PDE elements that follows.

We may introduce a constraint equation that specifies which of the node potentials in one element must equal a stated node potential in another. When this is done for all common nodes, a constraint matrix C may be assembled. For example, the structure depicted in Figure 9.2, when connected as indicated, results in

$$\begin{bmatrix} 1 & . & . & -1 & . & . & . & . & . \\ . & . & 1 & . & -1 & . & . & . & . \\ . & . & 1 & . & . & . & . & -1 & . \\ . & . & . & . & 1 & . & . & -1 & . \\ . & . & . & . & . & 1 & -1 & . & . \end{bmatrix} \begin{bmatrix} \phi_1 \\ \phi_2 \\ \phi_3 \\ \phi_4 \\ \phi_5 \\ \phi_6 \\ \phi_7 \\ \phi_8 \\ \phi_9 \end{bmatrix} = \begin{bmatrix} 0 \\ 0 \\ 0 \\ 0 \\ 0 \\ 0 \\ 0 \\ 0 \\ 0 \end{bmatrix}$$

By combining all of the individual constraints, for any such system, we obtain a matrix constraint equation

$$C^T \begin{bmatrix} \phi_1 \\ \phi_2 \\ . \\ . \\ \phi_i \\ . \\ \phi_M \end{bmatrix} = 0 \tag{9.35}$$

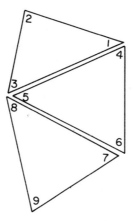

Figure 9.2   A disjoint assemblage of elements

where the superscript T denotes the transpose. Matrix C consists of ± 1 entries and many zeros. If this equation is satisfied, and if the elements are compatible, the potential will be continuous across interfaces between finite elements.

The complete matrix equation that must be solved for all node potentials may be generated by the method of Lagrange multipliers.[10,32] With the introduction of a vector of Lagrange multipliers λ (whose entries are related to interelement flux densities),

$$
\begin{bmatrix}
\boxed{S_1} & & 0 & \\
& \ddots & & C \\
0 & & \boxed{S_M} & \\
& C^T & & 0
\end{bmatrix}
\begin{bmatrix}
\phi_1 \\ \vdots \\ \vdots \\ \phi_M \\ \lambda
\end{bmatrix}
=
\begin{bmatrix}
b_1 \\ \vdots \\ \vdots \\ b_M \\ 0
\end{bmatrix}
\qquad (9.36)
$$

Equation (9.36) is a <u>diakoptical system</u> for which a convenient solution procedure exists.[32] It is a diakoptical system comprising a number of pre- viously discretized subsystems $[S_i]$ and has the same form as that resulting from hybrid analysis of an electrical network.[32] It is not of the same form as the discretization of a diakoptical system of a number of continuum subsystems.

Note that in order to eliminate variables we have *increased* the order of the matrix system. Almost invariably, in such PDE problems, one would define nodes to be common *a priori* and so reduce, rather than increase, the order. Lagrange multipliers are rarely used in practical applications involving com- patible PDE elements. However, understanding the technique is important for the methods that follow.

In the foregoing explanation, submatrices ϕ_i and $[S_i]$ have been ascribed to an element. However, they could equally pertain to an assemblage of elements wherein common nodes are joined together. Such an element structure can then be considered as a single finite element.

9.2.2 PDE formulations with dirichlet boundary condition constraints

In this section, it is shown how clusters of finite elements in free space may be viewed as sets of linear equation subsystems with the interaction effects furnished by constraint equations resulting from equivalent source distributions.

Consider the problem depicted in Figure 9.3. The region is unbounded with local areas of nonhomogeneous material (e.g. inhomogeneous, anisotropic dielectric objects). Sources or boundaries subject to Dirichlet conditions might be specified. Neither the IE nor the PDE formulation is by itself easily applied, but a combination of these techniques is effective in solving the field problem.

The finite region is segmented as follows: each region of nonhomogeneous material is enclosed in a 'picture-frame' leaving an infinitely large free-space region external to these 'picture-frames', as shown in Figure 9.4. The PDE

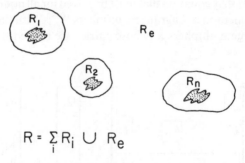

Figure 9.3 Picture-frame segmentation of a bounded region R

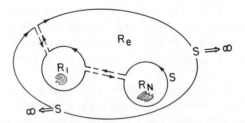

Figure 9.4 The region R_e external to the picture-frames R_i

formulation is used within each picture-frame and an IE formulation for the unbounded free-space region. Simultaneous solution of the two formulations produces the solution to the given problem.

The PDE formulation within each picture-frame is completely standard. In fact, as each picture-frame is as yet isolated from all the others, the matrix equation produced using the PDE finite-element formulation is of the same structure as that given in (9.34) with $S_i \boldsymbol{\phi}_i = \mathbf{b}_i$ as the matrix equation for the region within each picture-frame. The region within each picture-frame can consist of a single element or an assemblage of elements. Just as the finite elements in the previous case were mutually constrained to enforce continuity, as in (9.36), the picture-frames here are mutually constrained to ensure that the open-region problem is solved, and an integral equation formulation is used to construct the constraint.

The appropriate integral expression for a bounded region R with a boundary S is[16]

$$\phi(\mathbf{r}) = \gamma \oint_S \left(G(\mathbf{r}|\mathbf{r}') \frac{\partial \phi(\mathbf{r}')}{\partial n'} - \phi(\mathbf{r}') \frac{\partial G(\mathbf{r}|\mathbf{r}')}{\partial n'} \right) ds' \tag{9.37}$$

where
$$\gamma = \begin{cases} 0 & \mathbf{r} \text{ not in } R \\ 1 & \mathbf{r} \text{ in } R \\ 2 & \mathbf{r} \text{ on smooth part of } S. \end{cases} \tag{9.38}$$

This equation relates potential at a point $\mathbf{r}$ to potential and its derivatives on the boundary S through an integration involving the appropriate freespace Green's function.

By drawing R and S as shown in Figure 9.4, and taking part of S to infinity (over which the integral can be shown to vanish), one arrives at the integral expression for the R_e external to the picture frames. If ϕ and $\partial \phi / \partial n$ on S satisfy (9.37) (with $\mathbf{r}$ on S as well) the potential everywhere in R_e satisfies the PDE exactly. Thus (9.37) may be used to produce the necessary constraint on the PDE picture-frame solutions, allowing the correct solution of the open-region field problem to be obtained. As the integration in (9.37) is only required on picture-frame boundaries, the picture-frames may be as far apart as we like: the 'action-at-a-distance' property of the Green's function links the solutions together through the integration.

The full details of this numerical technique are given elsewhere.[2] However, the essential algorithmic principles follow using a notation slightly different from that used in the above-mentioned paper.

Referring to Figure 9.5, the statement of the problem, at first sight, appears to be incomplete, from a PDE viewpoint, as the potentials at the boundary nodes $\boldsymbol{\phi}_B$ are not given. The extra information required is furnished by a set of constraints developed by application of (9.37).

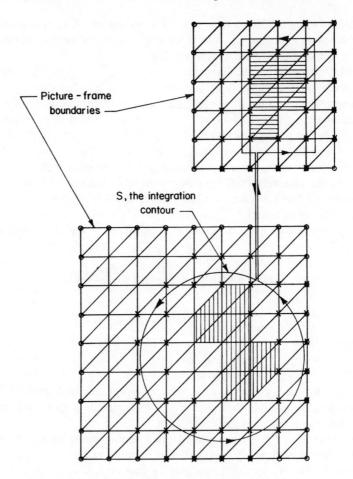

Figure 9.5 A specific example employing nonvariational integral constraints. x denotes nodes belonging to ϕ_S; o denotes nodes belonging to ϕ_B

If **r** in (9.37) is on the boundary then the Green's function becomes singular. However, this difficulty may be quite literally sidestepped by defining an integration contour S (as in Figure 9.5) that does not coincide with the previous one upon which $\phi(\mathbf{r})$ is evaluated. The integration, defined by (9.37), is performed over the new contour S with the restriction that **r** and **r'** should not coincide with the exception of cuts linking regions. The spacing between S and the picture-frame boundary, which must contain a homogeneous medium, guarantees that the Green's function is regular. Note that the use of two contours is a convenient feature of this method but is not essential.

The integral constraint is obtained by evaluating (9.37), for **r** at each node point along the boundary in turn, by integration over S. ϕ and $\partial\phi/\partial n$ are expressible in terms of node-point potentials within elements through which S passes. Using (9.33), disregarding the distant integration which vanishes, (9.37) becomes

$$\phi(\mathbf{r}) = \int_S \left(G(\mathbf{r}|\mathbf{r}') \frac{\partial \boldsymbol{\alpha}_S^T(\mathbf{r}')}{\partial n'} - \boldsymbol{\alpha}_S^T(r') \frac{\partial G(\mathbf{r}|\mathbf{r}')}{\partial n'} \right) ds' \boldsymbol{\phi}_S \qquad (9.39)$$

within the region R.

$\boldsymbol{\phi}_S$ is a column matrix containing node-point potentials within finite elements through which S passes. $\boldsymbol{\alpha}_S$ is a column matrix containing the corresponding interpolation functions. Evaluation of (9.39) at each boundary-node point produces the following matrix equation:

$$\boldsymbol{\phi}_B = \boldsymbol{P}\boldsymbol{\phi}_S. \qquad (9.40)$$

$\boldsymbol{P}$ has as many rows as there are boundary nodes and as many columns as there are nodes in the elements through which S passes. Specifically, there are 52 and 64, corresponding to the problem in Figure 9.5, out of a total of 117 nodes. It is clear that (9.40) results from the application of the method of point matching. It is a special case of the method of weighted residuals in which the expansion functions are contained within the column matrix $\boldsymbol{\alpha}_S$ and the testing functions are Dirac delta functions located at boundary-node points.

As $\boldsymbol{\phi}_B$ and $\boldsymbol{\phi}_S$ generally contain fewer entries than $\boldsymbol{\phi}$, they may be expressed, in terms of $\boldsymbol{\phi}$, through the matrices $\boldsymbol{Q}$ and $\boldsymbol{R}$ in the following way:

$$\boldsymbol{\phi}_B = \boldsymbol{Q}\boldsymbol{\phi} \qquad \boldsymbol{\phi}_S = \boldsymbol{R}\boldsymbol{\phi}. \qquad (9.41)$$

$\boldsymbol{Q}$ and $\boldsymbol{R}$ contain a one in each row and, in general, many columns of zeros. Upon substitution of (9.41) into (9.40) followed by rearrangement, one obtains the matrix constraint equation

$$\boldsymbol{C}^T \boldsymbol{\phi} = \boldsymbol{0}. \qquad (9.42)$$

Considering (9.34) to refer to any cluster of finite elements, the constraints may be incorporated by a Lagrange-multiplier technique similar to (9.36). But, as is seen in Section 9.2.4, the formulation is not symmetrical. Another distinction is that the nonzero terms in $\boldsymbol{C}$ are not simply $+1$ and -1 entries: they will be floating-point numbers and, for the Helmholtz equation, complex numbers as well. In addition, this matrix constraint equation has more nonzero entries than that of the previous section. In the same way that the previous constraint equations linked the individual finite elements (or sets of finite elements) to one another, these constraint equations match the picture-frames to each other and to the free-space solution in the unbounded region.

Solution of the resulting set of linear equations produces the node-point potentials $\boldsymbol{\phi}$ and thus the field patterns within each picture frame are immediately available. In a sense, then, we obtain a set of snapshots showing local parts of a field that extends over our infinite region. By concentrating on regions of interest, considerable economies are realized.

An example of a two-dimensional, time-harmonic problem[2] is shown in Figure 9.6. The diagram shows one-quarter of the field pattern resulting from symmetrical scattering of a wave from two dielectric obstacles. Note how the

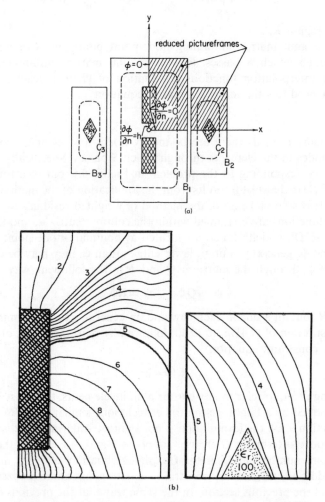

Figure 9.6 Antenna-type problem with dielectric obstacles using three picture frames. (a) Cross section showing the picture frames; (b) harmonic solution, contours, proportional to H_z, at 5 GHz, at phase zero, in the reduced picture frames. (From *IEEE Transactions on Microwave Theory and Techniques*, **MTT-20**, No. 12, 846, December 1972. Reproduced by permission of IEEE)

contours jump across the intervening space. The saving in elements and associated computational costs, due to economies realized in the intervening space, can be considerable.

Each formulation (the PDE and the IE) is used to its best advantage, only the simple free-space Green's functions are required, and the Green's function singularities are 'sidestepped' during computation. Although it is computationally very convenient, this scheme suffers from two disadvantages: the free-space region between S and each picture-frame, required for the double integration–contour algorithm, may require the picture-frames to be larger than we really need; and because the constraint equations result from a point-matching moment discretization of an integral equation, we have no guarantee that the overall solution will converge nor will the overall accuracy be as high as would be obtained by a variational scheme. Moreover, the accuracy may depend rather critically upon the location of the testing points.[24]

Another implementation of the moment method is described by Mei.[26,27] Comparison with other, fundamentally similar, schemes is made elsewhere.[2]

9.2.3 Finite elements for IE problems

In Reference 4 it was shown that strict application of the variational method, using the form of (9.14) for real operator and functions, could be used to solve certain integral equation problems. Specifically the electrostatic problem, with unknown charge distributions and specified potential distributions in free space, was solved although the Helmholtz problem is amenable to the same method. The paper then described a technique for handling the interface problem (specifically, that between air and a dielectric medium) which caused the introduction of nonself-adjointness. The use of the adjoint operators in order to deal with interfaces in a more efficient fashion is described in Reference 19. It is evident that the latter approach to the interface problem corresponds to the application of the Galerkin method, as was pointed out previously following Equation (9.30).

The foregoing references describe the use of free-space Green's functions, rather than special Green's functions, for the solution of general boundary shapes involving various boundary conditions.

In order to obtain a source distribution over a surface S which is at a specified potential, S is segmented into M finite elements. (The extension to problems involving interfaces is fairly easily made.[19]) Then, by analogy with the PDE case, the source distribution is written as

$$\sigma(\mathbf{r}) = \mathbf{\alpha}_i^T \mathbf{\sigma}_i \qquad \mathbf{r} \text{ on element } i \qquad (9.43)$$

where $\mathbf{\sigma}_i$ is a vector of source values at nodes in element i and $\mathbf{\alpha}_i$ is a vector of associated interpolatory functions, e.g. pulses, polynomials, etc.

Equations (9.20)–(9.22) define the appropriate matrix system. Using $\mathbf{\sigma}$ for v

and $g(\mathbf{r})$ for $f(\mathbf{r})$, we obtain

$$
\begin{bmatrix}
D_{11} & D_{12} & . & . & . & D_{1M} \\
D_{21} & & & & & . \\
. & & & & & . \\
. & & & & & . \\
D_{M1} & . & . & . & . & D_{MM}
\end{bmatrix}
\begin{bmatrix}
\sigma_1 \\
. \\
. \\
. \\
\sigma_M
\end{bmatrix}
=
\begin{bmatrix}
\boldsymbol{b}_1 \\
. \\
. \\
. \\
\boldsymbol{b}_M
\end{bmatrix}
\tag{9.44}
$$

where each entry in the square matrix is generally a submatrix

$$
D_{ij} = \frac{1}{\epsilon_0} \int_{S_i} \int_{S_j} \boldsymbol{\alpha}_i(\mathbf{r}) \boldsymbol{\alpha}_j^T(\mathbf{r}') G(\mathbf{r}|\mathbf{r}') \mathrm{d}s' \, \mathrm{d}s \tag{9.45}
$$

and each

$$
\boldsymbol{b}_i = \int_{S_i} \boldsymbol{\alpha}_i(\mathbf{r}) g_i(\mathbf{r}) \mathrm{d}s \tag{9.46}
$$

is generally a subvector. $g_i(r)$ is the Dirichlet condition on S_i, the ith finite element on S.

The off-diagonal matrix entries D_{ij}, $i \neq j$, represent the mutual constraints of the integral equation finite elements. It may be observed that the coefficient matrix is symmetric and, as potential at any point is due to the cumulative effect of all sources in the system, the matrix is also dense.

Integrations required in generating (9.44) involve singularity of the Green's function whenever $\mathbf{r} = \mathbf{r}'$. The addition–subtraction technique,[4,21] which is a useful procedure for computing the matrix entries, is based upon the following identity:

$$
\int_{S_i} \boldsymbol{\alpha}_i^T(\mathbf{r}') G(\mathbf{r}|\mathbf{r}') \mathrm{d}s' = \int_{S_i} [\boldsymbol{\alpha}_i^T(\mathbf{r}') G(\mathbf{r}|\mathbf{r}') - \boldsymbol{\alpha}_i^T(\mathbf{r}) G_s(\mathbf{r}|\mathbf{r}')] \mathrm{d}s'
$$
$$
+ \int_{S_i} \boldsymbol{\alpha}_i^T(\mathbf{r}) G_s(\mathbf{r}|\mathbf{r}') \mathrm{d}s' \tag{9.47}
$$

where $\mathbf{r}$ is on S_i. The function $G_s(\mathbf{r}|\mathbf{r}')$ is chosen so that the integrand of the first integral is regular when $\mathbf{r} = \mathbf{r}'$ and the second integral may be obtained analytically. In order to perform the second integration analytically, flat elements have been used. Extension to curved isoparametric surfaces is reported elsewhere.[21,22]

A very simple example, which shows the importance of using polynomials instead of the frequently used pulse functions, is indicated in the results of Figure 9.7. Better convergence of the polynomial approximation to that of the step approximation is evident. The extremely rapid convergence of the pulse trial function augmented by the addition of the singular edge function is also apparent.

An experiment, using isoparametric finite elements for integral equations,

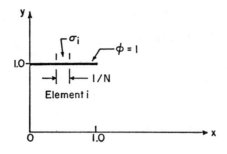

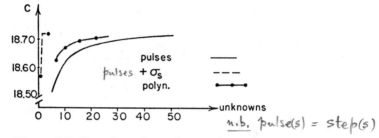

Figure 9.7 Boundary integral equation, variational computation of the capacitance (pF per metre depth) of two infinitely long parallel plates. Capacitance as a function of number of variables using pulse (with and without edge singular function) and polynomial expansion functions

was performed to compare the variational and point-matching methods.[21] The variational approach required about one-half the number of linear equations for the same accuracy. The boundary integral method (abbreviated BIM) is currently receiving considerable attention by engineers concerned with problems in elasticity.[21,28]

9.2.4 Mutual constraint of PDE and IE variational forms

The essence of a variational discretization[19] is that the expansion and testing function sets be the same and that the inner product of those functions be taken over the stated domain of definition of the functions. In solving a problem such as the one depicted in Figure 9.5, the expansion and testing functions were dissimilar. Consequently, the integral equation furnished a set of nonvariational constraints.

Let us consider the solution of an open-region problem, involving the simultaneous, variational solution of both a PDE and an IE. Figure 9.1 describes this problem, the surface S possibly being chosen arbitrarily. Using a variational, finite-element scheme, the interior PDE can be solved by first making a guess at a Neumann boundary condition $\left[\dfrac{\partial \phi}{\partial n}\bigg|_s = h(s) \right]$. We can then evaluate the potential distribution $\phi(s)$ along the boundary. Due to the fact that $h(s)$ was itself an estimate, $\phi(s)$ will only be approximate.

Having a known $\phi(s)$ estimate, (9.37) reduces to the form of (9.9) with $\partial \phi(\mathbf{r}')/\partial n'$ replacing $\sigma(\mathbf{r}')$. This may be solved variationally, as previously described, for $\partial \phi/\partial n$ thus producing (if this iterative procedure is convergent) an improved estimate to the PDE Neumann boundary condition. Note that this procedure applies continuity of potential and its normal derivative in sequence.

As an alternative to solving the coupled PDE–IE problem iteratively, as just described, the problem may be formulated as a set of IE-generated constraints upon a PDE system thus permitting solution in a direct fashion.

To produce a set of constraints, we begin with (9.37) in which $\mathbf{r}$ and $\mathbf{r}'$ are taken along the same boundary S. Therefore $\gamma = 2$ and so

$$\int_S \frac{\partial \phi}{\partial n'} G(\mathbf{r}|\mathbf{r}')\mathrm{d}s' - \int_S \phi(\mathbf{r}')\frac{\partial G}{\partial n'}(\mathbf{r}|\mathbf{r}')\mathrm{d}s' = \tfrac{1}{2}\phi(\mathbf{r}). \tag{9.48}$$

Using (9.33), one obtains

$$\left(\int_S \frac{\partial \boldsymbol{\alpha}(r')^T}{\partial n'} G(\mathbf{r}|\mathbf{r}')\mathrm{d}s' - \int_S \boldsymbol{\alpha}(\mathbf{r}')^T\frac{\partial G}{\partial n'}(\mathbf{r}|\mathbf{r}')\mathrm{d}s' - \tfrac{1}{2}\boldsymbol{\alpha}(\mathbf{r})^T \right)\boldsymbol{\phi} = 0. \tag{9.49}$$

Premultiplying $\partial \boldsymbol{\alpha}(\mathbf{r})/\partial n$ and integrating over S yields

$$\left(\int_S\int_S \frac{\partial \boldsymbol{\alpha}(\mathbf{r})}{\partial n}\frac{\partial \boldsymbol{\alpha}(\mathbf{r}')^T}{\partial n} G(\mathbf{r}|\mathbf{r}')\mathrm{d}s'\mathrm{d}s - \int_S\int_S \frac{\partial \boldsymbol{\alpha}(\mathbf{r})}{\partial n}\boldsymbol{\alpha}(\mathbf{r}')^T\frac{\partial G}{\partial n}(\mathbf{r}|\mathbf{r}')\mathrm{d}s'\mathrm{d}s \right.$$
$$\left. - \tfrac{1}{2}\int_S \frac{\partial \boldsymbol{\alpha}(\mathbf{r})}{\partial n}\boldsymbol{\alpha}(\mathbf{r})^T\mathrm{d}s \right)\boldsymbol{\phi} = 0 \tag{9.50}$$

in which care must be exercised in handling the Green's function singularities. Equation (9.50) may be written in the form

$$\boldsymbol{C}^T\boldsymbol{\phi} = 0. \tag{9.51}$$

This set of constraints must augment a partial differential equation formulation set of equations. Note that, in (9.50), it is only necessary to perform integrations involving interpolatory polynomials within elements adjacent to the boundary. Thus $\boldsymbol{C}$ has relatively few columns and is usually sparse.

There is an alternative formulation that raises some fundamental questions. Were we to have premultiplied (9.49) by $\alpha(\mathbf{r})$ and then integrated it, the appropriate system of constraint equations, obtained using this Galerkin approach,[19] would be

$$\left(\int_S \int_S (G(\mathbf{r}|\mathbf{r}') \; \boldsymbol{\phi}(\mathbf{r}) \frac{\partial \boldsymbol{\alpha}^T(\mathbf{r}')}{\partial n'} - \boldsymbol{\alpha}(\mathbf{r}) \; \boldsymbol{\alpha}(\mathbf{r}')^T \frac{\partial G}{\partial n'}(\mathbf{r}|\mathbf{r}') ds' ds \right.$$
$$\left. - \frac{1}{2} \int_S \boldsymbol{\alpha}(\mathbf{r}) \; \boldsymbol{\alpha}^T(\mathbf{r}) ds \right) \boldsymbol{\phi} = 0 \tag{9.52}$$

It would be very useful to have some numerical comparisons between the use of (9.50) and (9.52).

Another approach, involving mutual constraint of PDE and IE variational forms, is used in the pin insulator example that is discussed later. The technique is as follows.

Let us assume that field in the (possibly inhomogeneous) interior region is posed as a problem with an unknown Neumann boundary condition around the picture frame. The appropriate functional is therefore[25]

$$F = \int_R \epsilon \nabla \phi \cdot \nabla \phi d\Omega - 2 \oint_S \phi h ds \tag{9.53}$$

with

$$\epsilon \frac{\partial \phi(s)}{\partial n} = h(s) \tag{9.54}$$

We consider that $h(s)$ is fixed although unknown within the context of the problem contained in the picture frame.

The exterior problem is governed by (9.37) with $\gamma = 2$ for $\bar{r} = s$. For convenience, let

$$\lambda(s) = \frac{\partial \phi(s)}{\partial n} \tag{9.55}$$

for the homogeneous exterior region and so

$$\lambda(s) = -\epsilon_r \frac{\partial \phi(s)}{\partial n} \tag{9.56}$$

where the right-hand side derivative in (9.56) pertains to the interior region. The negative sign results from a change in direction of the unit normal, the normal being taken in the direction outward from the region under consideration.

Substituting (9.56) into (9.53),

$$F = \int_R \epsilon_r \nabla \phi \cdot \nabla \phi d\Omega + 2 \int_S \phi \lambda ds \tag{9.57}$$

in which ϵ_r is the interior to exterior permittivity ratio and F is suitably amended. Taking

$$\lambda(s) = \mathbf{\alpha}(s)^T \mathbf{\lambda} \tag{9.58}$$

and

$$\phi(s) = \mathbf{\alpha}(s)^T \mathbf{\phi} \tag{9.59}$$

we obtain

$$F = \mathbf{\phi}^T \int_R \epsilon_r \nabla \mathbf{\alpha}(s) \cdot \nabla \mathbf{\alpha}(s)^T d\Omega \mathbf{\phi} + 2\mathbf{\phi}^T \int_S \mathbf{\alpha}(s)\mathbf{\alpha}(s)^T ds \mathbf{\lambda} \tag{9.60}$$

Differentiating,

$$\frac{\partial F}{\partial \mathbf{\phi}^T} = 2 \int_R \epsilon_r \nabla \mathbf{\alpha}(s) \cdot \nabla \mathbf{\alpha}(s)^T d\Omega \mathbf{\phi} + 2 \int_S \mathbf{\alpha}(s)\mathbf{\alpha}(s)^T ds \mathbf{\lambda} \tag{9.61}$$

Thus

$$S\mathbf{\phi} + G\mathbf{\lambda} = \mathbf{b} \tag{9.62}$$

where

$$S = \left[\int_R \epsilon_r \nabla \alpha_i(s) \cdot \nabla \alpha_j(s) d\Omega \right] \tag{9.63}$$

$$G = \left[2 \oint_S \alpha_i(s)\alpha_j(s) ds \right] \tag{9.64}$$

and $\mathbf{b}$ has been added to account for PDE source terms.

From (9.37),

$$\phi(s) + 2 \oint_S \phi(s') \frac{\partial G(s|s')}{\partial n'} ds' - 2 \oint_S G(s|s')\lambda(s') ds' = 0 \tag{9.65}$$

Substituting (9.58) and (9.59),

$$\left[\mathbf{\alpha}(s)^T + 2 \oint_S \mathbf{\alpha}(s')^T \frac{\partial G(s|s')}{\partial n} ds' \right] \mathbf{\phi} - 2 \oint_S G(s|s')\mathbf{\alpha}(s')^T ds' \mathbf{\lambda} = 0 \tag{9.66}$$

and therefore

$$\left[\oint_S \mathbf{\alpha}(s)\mathbf{\alpha}(s)^T ds + 2 \oint_S \oint_S \mathbf{\alpha}(s)\mathbf{\alpha}(s')^T \frac{\partial G(s|s')}{\partial n} ds' ds \right] \mathbf{\phi}$$
$$- 2 \oint_S \oint_S G(s|s')\mathbf{\alpha}(s)\mathbf{\alpha}(s')^T ds' ds \mathbf{\lambda} = 0. \tag{9.67}$$

Thus

$$C^T \phi - Z\lambda = 0 \tag{9.68}$$

in which

$$C^T = \left[\oint_S \alpha_i(s)\alpha_j(s)ds + 2\oint_S \int_S \alpha_i(s)\alpha_j(s') \frac{\partial G(s|s')}{\partial n'} ds'ds \right] \tag{9.69}$$

and

$$Z = \left[2\oint_S \int_S G(s|s')\alpha_i(s)\alpha_j(s')ds'ds \right]. \tag{9.70}$$

Equations (9.62) and (9.68) may be written in compound matrix form in the following way:

$$\begin{bmatrix} S & G \\ C^T & -Z \end{bmatrix} \begin{bmatrix} \phi \\ \lambda \end{bmatrix} = \begin{bmatrix} b \\ 0 \end{bmatrix} \tag{9.71}$$

If sources exist in the exterior region, it is easy to show that the bottom entry in the right-hand-side vector is nonzero. (Constraint equations (9.50) and (9.52) would produce the same matrix form but with Z a null matrix.) Matrix equation (9.71) is of diakoptical form[32] and may be very conveniently solved.

From (9.62),

$$\phi = S^{-1}b - S^{-1}G\lambda \tag{9.72}$$

Premultiplying by C^T and employing (9.68),

$$\lambda = (C^T S^{-1}G + Z)^{-1} C^T S^{-1} b \tag{9.73}$$

Substituting (9.73) into (9.72),

$$\phi = S^{-1}b - S^{-1}G(C^T S^{-1}G + Z)^{-1} C^T S^{-1}b \tag{9.74}$$

in which the first term on the right-hand side yields the solution that results from a homogeneous Neumann boundary condition (i.e. open circuit) at the picture frame and the second term alters the solution in order to add the 'loading' effect of the rest of space.

Although (9.74) appears to be rather complicated, it in fact describes a fairly straightforward algorithm. Note that if one includes all α_i in $\boldsymbol{\alpha}$ then C^T, Z and G^T are square and have as many rows and columns as S. However, all rows that do not correspond to nodes in elements adjacent to the picture-frame boundary are empty. In practice, therefore, C^T, G^T, and Z have relatively few nontrivial rows and columns. These matrices are quite sparse. Generally, S is also sparse and it ought not to be actually inverted as its inverse would be rather dense. It is preferable to use a sparse-matrix method to produce a set of factor matrices that are individually sparse.[32] For a given picture frame shape, and element edge partitioning and order along it, the G, C^T and Z matrices can be generated and stored to use for any interior configuration. The matrix multiplications are then mainly performed by doing a sequence of sparse matrix-by-vector multiplications except for the interior bracketed quantity which is of small order and can be inverted in its entirety.

Of particular note is that the matrices C^T, G, and Z involve interpolatory functions and not their derivatives. This means that, if the integration contour corresponds to an interface between elements, only the nodes and associated shape functions along the· integration contour are required. This happens because the normal derivative is treated as an independent variable. Contrast this with equations (9.39), (9.50), and (9.52) which require the involvement of nodes adjacent to as well as along the integration contour.

The analyses that led to (9.71) were based upon the assumption of an unknown Neumann boundary condition over S. As an alternative, an unknown Dirichlet boundary condition over S can be assumed.[2] The diakoptical formulation (due to B. W. Gawick) follows.

If S is within B, referring to Figure 9.5, then

$$\phi_S = C_1^T \phi_1 \tag{9.75}$$

in which ϕ_S was previously defined and ϕ_I is a column matrix containing all interior node potentials within and not upon S.

Finite element discretization of a closed region results in the system of equations

$$S\phi = b \tag{9.76}$$

in which b is due to sources and S is singular unless some reference potentials are imposed. If the boundary node potentials ϕ_B are considered fixed (although unknown) then a corresponding number of rows ought to be deleted from (9.76). This is because those rows were derived by treating boundary node potentials as variational parameters—a role which fixed potentials cannot play. Only those internal nodes contained within ϕ_I are variational parameters. As a result, we have

$$S_I \phi_I + G \phi_B = b \tag{9.77}$$

where b is appropriately redefined.

As one example of the Dirichlet formulation, application of (9.39) to each node on the picture-frame boundary in turn yields

$$\phi_B = C_I^{Td} \phi_I \tag{9.78}$$

if the contour S is within B and if all normal derivatives employ no nodes upon B.

Therefore,

$$\begin{bmatrix} S_I & G \\ C_I^T & -I \end{bmatrix} \begin{bmatrix} \phi_I \\ \phi_{B_.} \end{bmatrix} = \begin{bmatrix} b \\ 0 \end{bmatrix} \tag{9.79}$$

where I is the unit matrix. (In general, if picture-frame boundary nodes are involved in the normal derivatives, then $-I$ must be replaced by a general matrix C_B^T.) This system may be solved in the previously-described fashion, yielding

$$\phi_I = S^{-1}b - S^{-1}G(I + C^TS^{-1}G)^{-1}C^TS^{-1}b \tag{9.80}$$

A different sequence of eliminations produces alternative algorithms. For example, substituting ϕ_B from (9.78) into (9.77) produces

$$\phi_I = (S_I + GC_I^T)^{-1}b \tag{9.81}$$

for the Dirichlet formulation. For the Neumann formulation, substitution of λ from (9.68) into (9.62) produces

$$\phi = (S + GZ^{-1}C^T)^{-1}b \tag{9.82}$$

These formulations could be slightly more convenient to program. On the other hand, they lack some generality. For example, the previous schemes permit the linking together of several picture frames by an integral equation constraint.[2] (See Figure 9.6). Individual picture-frame locations can be changed or the contents of some of the picture frames can be altered without the necessity of again solving the entire matrix system.

A word or two regarding some variations on themes are in order.

In the special case of a Laplacian field, the total energy content over all space is known to be finite. Thus, the functional[7,8] may be evaluated over an external and internal region, i.e.

$$F = \int_{R_i} (\nabla\phi \cdot \nabla\phi - 2\phi\rho)d\Omega_i + \int_{R_e} \nabla\phi \cdot \nabla\phi \; d\Omega_e \tag{9.83}$$

in which R_i and R_e denote the interior and exterior regions respectively, the permittivity is taken to be constant throughout, and ρ is the source distribution which is taken to exist only within R_i. Considering the second integral,

$$\int_{R_e} \nabla\phi \cdot \nabla\phi \; d\Omega_e = -\int_{R_e} \phi\nabla^2\phi \; d\Omega_e - \oint_s \phi\frac{\partial\phi}{\partial n}ds \tag{9.84}$$

where n is the outward normal. If we are assured that all trial functions for the external field satisfy Laplace's equation, i.e. by invoking (9.37) as a constraint, then (9.83) becomes

$$F = \int_{R_j} (\nabla\phi \cdot \nabla\phi - 2\phi\rho)d\Omega_i - \oint_s \phi\frac{\partial\phi}{\partial n}ds. \tag{9.85}$$

The boundary integration in (9.85) causes a distinction to occur between the methods proposed here and those proposed by Williams and Cambrell[11] and earlier by Silvester and Hsieh.[29] The point is made by Csendes[30] with respect to the picture frame method.[2] As he observes, fields beyond the picture frame are not directly involved in the trial function energy minimization process of Reference 2. However, they are involved as constraints and so are, in fact, considered in the solution process. He also notes that a further distinction relates to the use of two concentric picture frames rather than one. This is not an essential distinction as a single frame could be used, as has already been pointed out.

It seems possible that the extra term in (9.85) could lead to higher accuracy in the solution of Laplace's equation. However, one of the references[29] presents an incomplete integral formulation (in its Equation (13)) but uses a variational procedure while the other[11] reports a point-matching (rather than variational) procedure used for applying the exterior constraint equation. Thus, it is rather difficult to make a sensible comparison with the procedures described here.

9.2.5   An illustrative example

Consider the pin insulator problem (suggested by L. M. Warren, University of Manchester Institute of Science and Technology) shown in Figure 9.8(a). Solution of the axisymmetric problem involves solution of

$$\frac{\partial}{\partial r}\left(r\epsilon\frac{\partial\phi}{\partial r}\right)+\frac{\partial}{\partial z}\left(r\epsilon\frac{\partial\phi}{\partial z}\right)=0. \tag{9.86}$$

Note that the form of (9.86) implies that although the permittivity is piecewise constant with position, the radial coordinate introduces an effective inhomogeneity, i.e. permittivity increasing linearly with r. MANFEP[15] caters for media having properties that vary continuously within elements (and for anisotropic media as well) and, therefore, specification of special axisymmetric elements is unnecessary. We thus use the partial differential equation formulation for the inhomogeneous region and the integral equation formulation for the exterior region. Note that it is necessary to use ring-type, axisymmetric Green's functions.[31]

Figure 9.8(b) shows the subdivision into 24 elements. At the time of this work, it was required that the outer boundary be polygonal, the algorithm not then being available for treating Green's function singularities along curved sides. Had curved outside surfaces been allowed[21,22] only 9 elements would have been required, all of them within the insulator.

The Dirichlet conditions $\phi=0$, $\phi=1$ are enforced as shown and the picture-frame is assumed to have a Neumann boundary condition. For a quadratic polynomial approximant, there are 63 node potentials to be found. The static

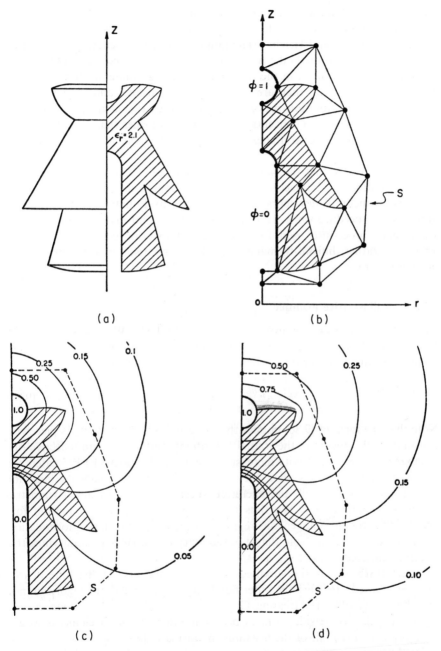

Figure 9.8 A variationally constrained example. (a) A high-voltage insulator in an open region; (b) subdivision into 24 elements; (c) voltage contours; (d) the effect of contamination

equipotentials are shown in Figure 9.8(c). Continuous and smooth ϕ is seen to result throughout.

A contaminated insulator is represented by permitting conducting material to reside on the surface of the insulator. Thus a portion of the interface is an equipotential. This is easily enforced by using Lagrange constraints, as in (9.36), to force node potentials on the contaminated surface to the same unknown value. For example, Figure 9.8(d) shows the effect due to contaminant at the top of the insulator. There is a considerable shift in the location of potential contours toward the ground.

9.3   CONCLUSIONS

The Galerkin/variational method is generally recommended because convergence is assured even for the general operator as long as it contains a positive-bounded-below component. For example, the Helmholtz operator, which is generally indefinite, contains the Laplacian operator and convergence is assured. From the integral equation viewpoint, the complex Helmholtz Green's function can be expanded in terms of a Taylor series. The first term is the self-adjoint Green's function for the Laplace equation and, therefore, convergence is assured. This assurance ought to be compared with, for example, the moment method using delta function testing which produces lesser accuracy and can produce results that are very sensitive with respect to testing locations. Therefore, of the various types of picture-frame methods available, the mutual constraint of PDE and IE variational forms is felt likely to produce the best results overall. Results derived from comprehensive experimentation, using a simple set of models, would be most useful for the guidance of those using these techniques. From a mathematical viewpoint, a recent work by Babuska[33] appears very relevant. Further study of the subject is necessary.

REFERENCES

1. Hammond, P. (1969). 'Sources and fields—some thoughts on teaching the principles of electromagnetism', *Int. Jnl. Elect. Eng.*, **7**, 65–76.   [Solton 1980]
2. McDonald, B. H. and Wexler, A. (1972). 'Finite element solution of unbounded field problems', *IEEE Trans. on Microwave Theory and Techniques*, **MTT-20**, No. 12, 841–847.
3. McDonald, B. H. Friedman, M., Decreton, M., and Wexler, A. (1973). 'Integral finite-element approach for solving the Laplace equation', *Electronics Letters*, **9**, No. 11, 242–244.
4. McDonald, B. H., Friedman, M., and Wexler, A. (1974). 'Variational solution of integral equations', *IEEE Trans. on Microwave Theory and Techniques*, **MTT-22**, No. 3, 237–248. (See also 5)
5. McDonald, B. H., Friedman, M., and Wexler, A. (1975). 'Correction to Variational solution of integral equations', '*IEEE Trans. on Microwave Theory and Techniques*, **MTT-23**, No. 2, 265–266.

6. Acton, F. S. (1970). *Numerical Methods That Work*, Harper and Row, New York.
7. Mikhlin, S. G. (1964). *Variational Methods in Mathematical Physics*, Macmillan, New York. pp. 313, 469–475.
8. Wexler, A. (1969). 'Computation of electromagnetic fields', *IEEE Trans. on Microwave Theory and Techniques*, **MTT-17**, No. 8, 416–439.
9. Zienkiewicz, O. C. (1971). *The Finite Element Method in Engineering Science*, McGraw-Hill, London.
10. Gallagher, R. H. (1975). *Finite Element Analysis*, Prentice-Hall, Englewood Cliffs, N. J.
11. Williams, C. G. and Cambrell, G. K. (1972). 'Efficient numerical solution of unbounded field problems', *Electronics Letters*, **8**, No. 9, 247–248.
12. Richards, D. J. and Wexler, A. (1972). 'Finite element solutions within curved boundaries', *IEEE Trans. on Microwave Theory and Techniques*, **MTT-20**, No. 10, 650–657.
13. Isaacson, E. and Keller, H. B. (1966). *Analysis of Numerical Methods*, Wiley, New York.
14. Cermak, I. (1969). 'Solution of Unbounded Field Problems by Boundary Relaxation.' *Ph.D. Thesis*, Department of Electrical Engineering, McGill University.
15. Friedman, M., Richards, D. J., Stevens, W. N. R., and Wexler, A. (1979). *MANFEP, User's Manual*, Report, Department of Electrical Engineering, University of Manitoba.
16. Burton, A. J. and Miller, G. F. (1971). 'The application of integral equation methods to the numerical solution of some exterior boundary value problems', *Proc. Royal Society*, **A323**, No. 1553, 201–210.
17. Cairo, L. and Kahan, T. (1965). *Variational Techniques in Electromagnetism*, Blackie, London.
18. Wexler, A. (1973). 'Finite element analysis of an inhomogeneous, anistropic, reluctance machine rotor', *IEEE Trans. on Power Apparatus and Systems*, **PAS-92**, 145–149.
19. Jeng, G. and Wexler, A. (1978). 'Self-adjoint variational formulation of problems having non-self-adjoint operators', *IEEE Trans. on Microwave Theory and Techniques*, **MTT-26**, No. 2, 91–94.
20. McDonald B. H. (1975). 'Constrained Variational Solution of Field Problems', *Ph.D. Thesis*, Department of Electrical Engineering, University of Manitoba,
21. Jeng, G. and Wexler, A. (1977). 'Isoparametric, Finite-Element, Variational Solution of Integral Equations for Three-Dimensional Fields', *Internat'l. J. Num. Meth. Eng.*, **11**, No. 9, 1455–1471. *Variational ~ Point matching*
22. Jeng, G. and Wexler, A. (1976). 'Finite Element Solution of Boundary Integral Equations', In A. Wexler (ed), *Proceedings of the International Symposium on Large Engineering Systems, August 9 to 12, 1976*, Pergamon Press, New York. 112–121.
23. Harrington, R. F. (1968). *Field Computation by Moment Methods*, Macmillan, New York.
24. Finlayson, B. A. (1972). *The Method of Weighted Residuals and Variational Principles*, Academic Press, New York and London.
25. Wexler, A. (1977). 'Isoparametric Finite-Element for Continuously Inhomogeneous and Anistropic Media', In C. A. Brebbia and W. G. Gray (eds), *Proceedings of the International Conference on Finite Elements in Water Resources, July 12 to 16, 1976*, Pentech Press, London, pp. 2.3–2.24.
26. Mei, K. K. (1974). 'Unimoment method of solving antenna and scattering Problems', *IEEE Trans. on Antennas and Propagation*, **AP-22**, No. 6, 760–766.
27. Chang, S-K. and Mei, K. K. (1976). 'Application of the unimoment method to

electromagnetic scattering of dielectric cylinders', *IEEE Trans. on Antennas and Propagation*, **AP-24**, No. 1, 35–42.

28. Cruse, T. A. and Rizzo, F. J. (eds) (1975). *Boundary-Integral Equation Method: Computational Applications in Applied Mechanics*, The American Society of Mechanical Engineers, New York. AMD-Vol. 11.

29. Silvester, P. and Hsieh, M.-S. (1971). 'Finite-element solution of 2-dimensional exterior-field problems', *Proc. IEE*, **118**, No. 12, 1743–1747.

30. Csendes, Z. J. (1976). 'A note on the finite-element solution of exterior-field problems', *IEEE Trans. on Microwave Theory and Techniques*, **MTT-24**, No. 7, 468–473.

31. Morse, P. and Feshbach, H. (1953). *Methods of Theoretical Physics*, McGraw-Hill, New York.

32. Wexler, A. (1979). 'Perspectives on the solution of simultaneous equations'. Report, Department of Electrical Engineering, University of Manitoba.

33. Babuška, I. (1973). 'The finite element method with Lagrangian multipliers', *Numer. Math.*, **20**, 179–192.

Finite Elements in Electrical and Magnetic Field Problems
Edited by M. V. K. Chari and P. P. Silvester
© 1980, John Wiley & Sons Ltd.

Chapter *10*

Applications of Integral Equation Methods to the Numerical Solution of Magnetostatic and Eddy-Current Problems

C. W. Trowbridge

10.1 INTRODUCTION

The work described in this chapter had as its stimulus the all-pervasive use of electromagnets in high energy physics and fusion research. From 1970 onwards there has been an increasing demand for genuine three-dimensional calculations in magnet design, and it was to meet this need that the GFUN 3-D program described in Section 10.3 was developed. The choice of the magnetization integral equation method was made in order to avoid the problem of developing global mesh generators and also to avoid the specification of external boundary conditions. Whilst the GFUN program has been successfully used for the design of many magnets with results corroborated by measurement, it is clear that for complicated geometry the program is expensive to use and there is a need to develop more economical algorithms. It was to try to satisfy this objective that the integral equation methods described in Sections 10.4 and 10.5 were developed.

With regard to time-dependent field problems, it has become increasingly more important to be able to assess the effects of eddy currents in the design of both accelerator and Tokamak magnets. Also there is considerable interest in this topic in the electric machines area and various methods of approach are currently under review.[16]

In Section 10.6 some preliminary results are given for the solution of eddy current problems using the integral equation method.

10.2 BASIC EQUATIONS FOR MAGNETOSTATICS

The equations describing magnetostatic fields are, in SI units,

$$\operatorname{curl} \mathbf{H} = \mathbf{J} \tag{10.1}$$

$$\text{div} \ \mathbf{B} = 0 \tag{10.2}$$

$$\mathbf{B} = \mu(\mathbf{H})\mathbf{H} \tag{10.3}$$

$$\mathbf{M} = \frac{1}{\mu_0}\mathbf{B} - \mathbf{H} = \chi(\mathbf{H})\mathbf{H} \tag{10.4}$$

where $\mathbf{H}$ and $\mathbf{B}$ are the magnetic field intensity and magnetic flux density at a point respectively and $\mathbf{J}$ is the current density. Equation (10.3) is the constitutive relation for an isotropic material of permeability μ, a given function of the field intensity. Equation (10.4) defines the magnetization $\mathbf{M}$ and susceptibility χ. Furthermore, if the field at a point is expressed as the sum of the field $\mathbf{H}_c$ from current sources and $\mathbf{H}_m$ the field from induced magnetization sources, such that

$$\mathbf{H} = \mathbf{H}_c + \mathbf{H}_m \tag{10.5}$$

$\mathbf{H}_m$ can be expressed as the gradient of a scalar potential ϕ given by the relation

$$\mathbf{H}_m = -\text{grad} \ \phi \tag{10.6}$$

and

$$\phi = \frac{1}{4\pi}\int_{V'} \mathbf{M}(\mathbf{r}') \cdot \text{grad}\left(\frac{1}{\mathbf{r} - \mathbf{r}'}\right) dV' \tag{10.7}$$

where $\mathbf{r}$ and r' are position vectors of the field and source points respectively. The integration in Equation (10.7) is to be taken over the volume of all iron regions present in the problem.

Thus the magnetic field can be expressed in terms of a differential equation in ϕ obtained by Equations (10.1), (10.2), (10.5), and (10.6), so that

$$\text{div}(\mu\nabla\phi) = \text{div}(\mu\mathbf{H}_c). \tag{10.8}$$

The potential is subject to the boundary condition $\phi \to 0$ at infinity and continuity conditions

$$\left.\begin{array}{r} \phi_1 = \phi_2 \\[1mm] \mu_1\left(\dfrac{\partial\phi}{\partial n}\right)_1 - \mu_2\left(\dfrac{\partial\phi}{\partial n}\right)_2 = (\mu_1 - \mu_2)H_{cn} \end{array}\right\} \tag{10.9}$$

at the interface between two regions of different materials. In Equation (10.9) H_{cn} is the normal component of field from the current sources at the boundary.

As an alternative to seeking solutions of Equations (10.8) and (10.9) directly by use of a finite element (difference) method based on the differential operator, it may be advantageous in some circumstances (e.g. open systems) to define the problem in terms of an integral equation either by use of Equations

(10.5) and (10.7) or by transforming Equation (10.8) by use of Green's theorems. In the next three sections methods using integral equations for the magnetostatic problem are described.

10.3 MAGNETIZATION INTEGRAL EQUATION

10.3.1 Formulation

The integral equation for the unknown magnetization $\mathbf{M}$ is obtained from equations (10.5) and (10.7), i.e.

$$\mathbf{M}(\mathbf{r}) = \chi(\mathbf{r})\left[\mathbf{H}_0(\mathbf{r}) - \frac{1}{4\pi}\,\mathrm{grad}\int_{V'}\mathbf{M}(\mathbf{r}')\cdot\mathrm{grad}\left(\frac{1}{|\mathbf{r}-\mathbf{r}'|}\right)\mathrm{d}V'\right]. \qquad (10.10)$$

This equation forms the basis of the algorithm used in the GFUN computer program[13] first developed at the Rutherford Laboratory in 1970 and with many subsequent improvements.[1]

The algorithm uses an element discretization with the iron regions sub-divided into N prisms or tetrahedra. Equation (10.10) is replaced by a matrix equation by defining a set of basis functions.[9]

$$\mathbf{M}(\mathbf{r}) \cong \sum_{i=1}^{N}\mathbf{M}_i\delta_{ij} \qquad (10.11)$$

where

$$\delta_{ij} = \begin{cases} 1 & j=i \\ 0 & j\neq i \end{cases} \qquad (10.12)$$

and the $\mathbf{M}_i$ are the magnetization values to be determined for each element, assumed to be constant over the volume of an element.

Substituting (10.11) into (10.10) and requiring the equation to be satisfied at the centroid of each element the following set of equations is obtained.

$$\mathbf{M}_i = \chi_i\left(\mathbf{H}_{ci} - \sum_{j=1}^{N}\mathbf{M}_j G_{ij}\right) \qquad (10.13)$$

or in canonical form

$$\sum_{j=1}^{N}\left(\frac{\delta_{ij}}{\chi_j} + G_{ij}\right)M_j = H_{ci} \qquad (10.14)$$

or

$$\mathbf{AM} = \mathbf{H}_{\mathrm{c}}$$

where each element of G is a dyadic given by

$$G = \frac{1}{4\pi} \mathbf{\nabla} \int_{\text{Element}} \mathbf{e} \cdot \mathbf{\nabla} \left(\frac{1}{|\mathbf{r} - \mathbf{r}'|} \right) dV' \tag{10.15}$$

and $\mathbf{e}$ is a unit vector parallel to $\mathbf{M}$.

For prisms, tetrahedra, and polyhedra the integrations can be carried out in closed form.[22,6] The field from the current sources is obtained by integrating the Biot–Savart Law relation over the volume of the conductors, i.e.

$$\mathbf{H}_c = \frac{1}{4\pi} \int_{V'} \frac{\mathbf{J} \times (\mathbf{r} - \mathbf{r}')}{|\mathbf{r} - \mathbf{r}'|^3} dV'. \tag{10.16}$$

In fact $\mathbf{H}_c$ is available in the GFUN program for a range of basic conductor elements including straight blocks and curved sectors and many higher order circuits using these as building blocks.[5]

10.3.2 Solving the equations

For most 3-D problems the matrix in Equation (10.14) is too large to be held in main memory and so is stored as a direct access disk dataset. Also since the matrix has no convenient properties such as symmetry or sparsity, the method used is Gaussian elimination with back substitution. The matrix is partitioned so that only a fraction of the whole matrix need be in main memory at any time.[14] For nonlinear problems, i.e. cases where χ is a function of the magnetizing field, the procedure adopted originally in the development of the program was to treat the system of equations as quasi-linear and use a simple iterative scheme,

$$\mathbf{M}^{k+1} = \mathbf{A}^{-1} (\mathbf{M}^k) \mathbf{H}_c. \tag{10.17}$$

It was found that this simple method does converge in almost all cases, albeit rather slowly; however both secant and Newton–Raphson schemes are currently under test for use with the direct access disk system.

10.3.3 Data input/output and graphics

Extensive use of interactive graphics techniques has been a principal feature of GFUN throughout the program's development. Users prepare and check data at graphics terminals[20,21] and examine results displayed as line graphs or contour maps. Data files are stored on disk accessible by any of the programs belonging to the GFUN magnet design package. Relatively simple magnets can be defined and executed on line enabling an expert designer to gain a measure of physical understanding of his problem—he can in fact combine his

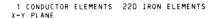

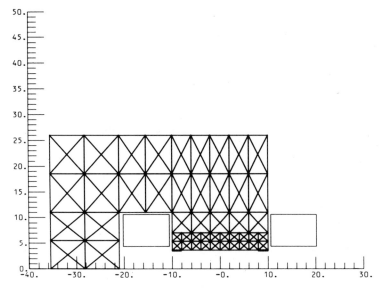

Figure 10.1 GFUN model of the EPIC C-magnet

experience with automatic optimization routines in order to achieve design problem solutions. For more complicated systems requiring many elements (>40 2-D or >27 3-D) the integral equation is solved by batch processing. Other facilities available for the analysis of results are field harmonics, forces, and particle tracking.

10.3.4 Results

10.3.4.1 *EPIC prototype dipole magnet*[2]

The cross section of this magnet is shown in Figure 10.1. The magnet length is 4.5 m, a distance very large compared to the gap between the pole faces and so a two-dimensional model was considered adequate. The design objective was to achieve a field homogeneity of 1 part in 10^4 over 100 mm pole width. The GFUN program was used to determine the sizes of the pole-edge shims both to compensate for the asymmetry caused by the C-shaped probe and to enlarge the good-field region. The prototype magnet was constructed and measured; Figure 10.2 shows a comparison between prediction and measurement.

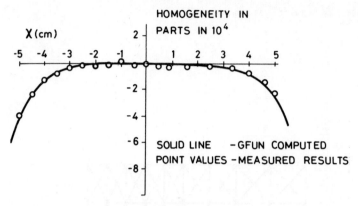

Figure 10.2   Comparison of measurements with computed values for the EPIC C-magnet. Solid line, GFUN computed; point values, measured results

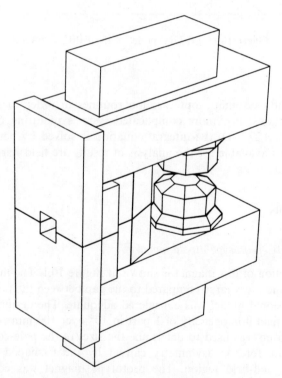

Figure 10.3   Computer-generated picture of the PEM magnet pole and yoke

10.3.4.2 PEM C-*magnet*

A computer-generated picture of this magnet, illustrating some of the advanced graphics techniques currently under development at the Rutherford Laboratory[14] is shown in Figure 10.3. The truncated conical pole tip is made from a special cobalt-steel; the cylindrical pole and return yokes are fabricated from various other magnet steels. The surface description of the model used to represent this magnet is shown in Figure 10.3: for example, it can be seen that the cylindrical parts are approximated by an octagonal prism. The rectangular channel through the back yoke was introduced to provide a low field path for a beam of charged particles to enter a polarized target placed in the good high field region under the pole tip. The element distribution used is shown in Figure 10.4.

The field distribution for this magnet has been measured[15] and these results are compared with the calculated fields for GFUN in Figure 10.5.

10.3.4.3 CERN *spectrometer model magnet*

A computer-generated picture of this magnet, designed as a 1/10 scale prototype for a forward spectrometer magnet, is shown in Figure 10.6. Field

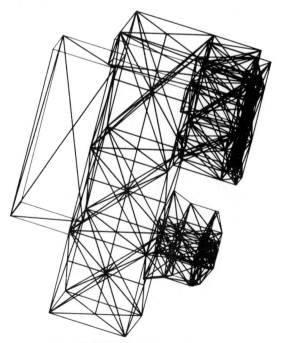

Figure 10.4 Tetrahedral mesh used in the PEM magnet computation

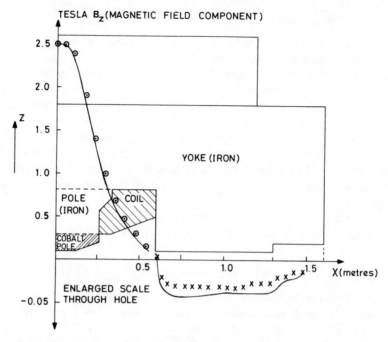

Figure 10.5 Comparison of measurements with computed values for the PEM magnet. Measured fields: ⊙ under pole; x through hole in back;——— calculated GFUN

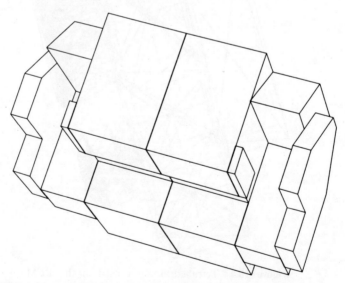

Figure 10.6 Computer-generated picture of the 1/10 scale model spectrometer magnet. (Only one quadrant shown)

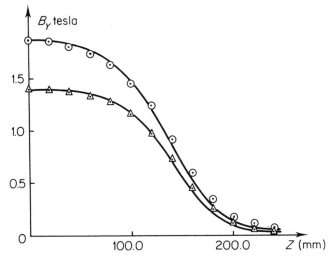

Figure 10.7 Comparison of measurements with computed values for the 1/10 scale model spectrometer magnet. ⊙ Measurements at 850 Amps, $Y=0.0$; △ measurements at 550 Amps, $Y=0.0$;———GFUN calculation

measurements[8] along the beam axis are compared with the calculated fields for GFUN in Figure 10.7.

These results show that a high precision (0.01%) on magnetic field prediction is possible for two-dimensional nonlinear problems providing accurate data of the material properties is known—this was the case for the EPIC dipole magnet. Clearly for three-dimensional magnets the same order of precision is not possible without a substantial increase in the number of elements used and a consequent enormous increase in computer resources. However, as the two cases presented here show, modest accuracy is attainable and may well be considerably better than indicated because of uncertainties in material properties. In the case of the PEM magnet the level of field in the hole was found to be particularly sensitive to the choice of material data.

Table 10.1 gives some statistics for this program run on an IBM 360/195 computer.

Table 10.1

Number of elements	Probable accuracy %	Magnetization		Fields at a point	
		Core (Kbytes)	CPU Time (s)	Core (Kbytes)	CPU Time (s)
32	10.0	210	20.0	210	0.30
100	5.0	720	360.0	210	1.00
200	1.0	720	1800.0	210	2.00

10.4 SCALAR POTENTIAL INTEGRAL EQUATION

10.4.1 Magnetistatic potential ϕ

The integral equation for the magnetostatic potential ϕ is obtained by substituting Equation (10.4)–(10.6) into (10.7):

$$\phi(\mathbf{r}) = -\frac{1}{4\pi} \int \chi(\mathbf{r}')(\mathbf{H}_c(\mathbf{r}') - \boldsymbol{\nabla}\phi(\mathbf{r}'))\frac{(\mathbf{r}'-\mathbf{r})}{|\mathbf{r}'-\mathbf{r}|^3}\,dV$$

$$(10.18)$$

An algorithm for the solution of Equation (10.18)[10] is described here. The iron is divided into tetrahedral elements and a set of either linear or quadratic

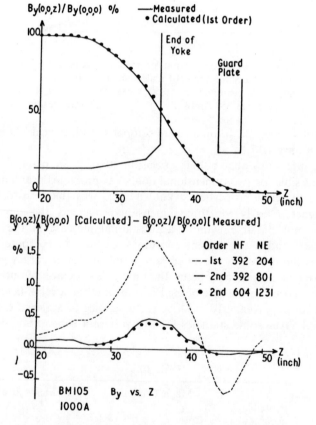

Figure 10.8 Comparison of measurements with computed values for the particle bending magnet BM105. The geometry is similar to the magnet shown in Figure 10.6

basis functions is set up to describe ϕ in each element:

$$\phi = \sum_{i=1}^{N} C_i \phi_i; \quad \nabla\phi = \sum_{1}^{N} C_1 \nabla\phi_i. \tag{10.19}$$

Here N is the number of nodes in the discretization, nodes being at the tetrahedron vertices, and in the quadratic case also midside positions. The C_i are the unknown values of ϕ at the nodes, and the ϕ_i are the known subsectional basis functions, that is they are linear or quadratic functions within one element and zero outside it. In order to evaluate the integrals in (10.18), H_c and χ are assumed to be constant or linear functions within each element for ϕ linear or quadratic respectively. Within these approximations the right-hand side of (10.18) is then determined in terms of C_i, yielding the equation:

$$f(\mathbf{r}) = 4\pi\phi + \int \chi(\mathbf{r}')(H_c(\mathbf{r}') - \nabla\phi(\mathbf{r})) \cdot \frac{(\mathbf{r}'-\mathbf{r})}{|\mathbf{r}'-\mathbf{r}|^3} dV' \neq 0. \tag{10.20}$$

The solution values C_i are chosen by making $f(\mathbf{r}) = 0$ at the N nodes. For the nonlinear problem these C_i are obtained by an iterative technique: if $f^{(k)}$ is the value of f at the kth iteration, $f^{(k+1)}$ is obtained by linearizing (10.20) in the neighbourhood of the previous approximation:

$$f^{(k)} + \frac{\partial f^{(k)}}{\partial C_i} \delta C_i = 0 \tag{10.21}$$

with $\partial f/\partial C_i$ obtained from (10.19) and (10.20)

$$\frac{\partial f}{\partial C_i} = 4\pi\phi_i - \int \nabla\phi_i \cdot \frac{(\mathbf{r}'-\mathbf{r})}{|\mathbf{r}'-\mathbf{r}|^3} dV$$

$$+ \int (H_c - C_i \nabla\phi_j) \cdot \frac{(\mathbf{r}'-\mathbf{r})}{|\mathbf{r}'-\mathbf{r}|} \frac{\partial\chi}{\partial H_m} (\nabla\phi_i)_m \tag{10.22}$$

This gives a set of equations for the changes δC_i in the C_i which can be expected to have quadratic convergence. The formula (10.22) allows for the possibility of anisotropic iron, if $\partial\chi/\partial H_m$ is not proportional to the mth component of the field H_m.

Iselin reports that the linear ϕ results compare poorly with GFUN, even when using a matrix of the same size, which permits 3 times as many nodes (the principal advantage of the scalar potential formulation). The potential values within the iron are sensible, but near the surface they oscillate violently. For quadratic ϕ the accuracy is comparable with GFUN, but the other potential advantage of the method, obtaining smoothly varying fields within the iron, is not yet realized.

10.4.2   Total scalar potential ψ

More recent work carried out at the Rutherford Laboratory[23,24] has shown that the use of scalar potential ϕ defined by Equation (10.6) is numerically unsatisfactory because of the cancellation between source and induced fields that can occur inside iron regions. This difficulty may be avoided entirely by the reformulating in terms of the total scalar potential ψ defined by:

$$\mathbf{H} = -\operatorname{grad}\,\psi \tag{10.23}$$

with ψ related to ϕ by

$$\psi = \phi + \psi_C \tag{10.24}$$

In which ψ_C is readily calculated from the prescribed conductors. The integral equation for ψ can be derived from Equations (10.7), (10.4) and (10.23) to obtain:[24]

$$\mu\psi(\mathbf{r}) = \frac{1}{4\pi}\int_{S'}\chi(\mathbf{r}')\psi(\mathbf{r}')\nabla\left(\frac{1}{|\mathbf{r}-\mathbf{r}'|}\right)\cdot d\mathbf{S}'$$

$$+\frac{1}{4\pi}\int_{V'}\psi(\mathbf{r}')\nabla\chi(\mathbf{r}')\cdot\nabla\left(\frac{1}{|\mathbf{r}-\mathbf{r}'|}\right)\cdot dV' + \psi_C(\mathbf{r}) \tag{10.25}$$

This equation is valid for a current free domain V' bounded by a surface S' situated in a prescribed external field. Equation (10.25) has been solved using both linear and quadratic base functions and the experience so far obtained indicates that the total scalar potential is very well behaved numerically.[24] In Figure 10.8 some preliminary results are presented for the end region of a particle bending magnet.

10.5   BOUNDARY INTEGRAL METHOD

10.5.1   Formulation

An alternative approach is a method of solution based on the integral equation for the magnetic scalar potential derived from Green's theorems.[19] This technique has been used previously for the solution of linear flow and elasticity problems[11,7,12] and has been applied to magnetostatic problems at the Rutherford Laboratory.[17] For linear problems, i.e. constant permeability, it is only necessary to define the boundaries of regions with different permeability, together with a far field boundary condition; the far field boundary can, however, be expanded to infinity. A region may consist of several surfaces that do not touch or intersect and this fact together with the use of symmetry allows the calculation of fields with minimal effort. It is also possible to extend the method to include nonlinear regions.

Green's second theorem[18] can be used to relate the potential ϕ at a point $\mathbf{r}$

inside a region to values of ϕ and $\partial\phi/\partial n$ on the surface. Hence

$$\phi(\mathbf{r}) = \frac{1}{4\pi} \int_{V'} \frac{1}{|\mathbf{r}-\mathbf{r}'|} \nabla^2 \phi(\mathbf{r}') dV' + \frac{1}{4\pi} \int_{S} \frac{1}{|\mathbf{r}-\mathbf{r}'|} \frac{\partial\phi(\mathbf{r}')}{\partial n} dS$$

$$- \frac{1}{4\pi} \int_{S} \phi(\mathbf{r}') \frac{\partial}{\partial n} \left(\frac{1}{|\mathbf{r}-\mathbf{r}'|} \right) dS. \tag{10.26}$$

If the permeability of a region is constant, then from Equation (10.8) the first integral in Equation (10.26) is zero. A possible means of including the effect of nonlinearity is to treat this term as a perturbation. Physically the kernels in the surface integrals of Equation (10.26) can be interpreted as single and double layer charges respectively or mathematically as singular solutions of Laplace's equation. Numerically, if the surface is discretized into m small areas over which ϕ and $d\phi/dn$ are constant, Equation (10.26) becomes

$$\phi_i = \frac{1}{4\pi} \sum_{j}^{m} \left(\frac{\partial\phi_j}{\partial n_j} \int_{S_j} \frac{dS_i}{|\mathbf{r}_i-\mathbf{r}_j|} - \phi_j \int_{S_j} \frac{\partial}{\partial n_j} \left(\frac{1}{|\mathbf{r}_i-\mathbf{r}_j|} \right) dS_j \right) \tag{10.27a}$$

or

$$\phi_i = \sum_{j}^{m} \left(A_{ij} \frac{\partial\phi_j}{\partial n_j} - B_{ij}\phi_j \right) \tag{10.27b}$$

Since either ϕ_j or $\partial\phi/\partial n_j$ is known for each boundary element for a well posed problem the unknown ϕ or $\partial\phi/\partial n$ can be determined by solving the set of linear Equations (10.27b). For the simple subsectional basis function chosen here the integrals in (10.27) are expressible in closed form.

A more interesting problem is that of regions with different permeabilities where there are interfaces between regions. In this case the interface between the regions is discretized and Equation (10.27a) is applied to the surfaces of both regions. Since both ϕ and $\partial\phi/\partial n$ are unknown along the interface the set of equations (10.27b) is underdetermined. Two extra equations must be introduced for each interface element and these equations are obtained directly from the continuity equation (10.9). The same ideas can be applied to problems consisting of any number of regions. It should be noted that the field $\mathbf{H}_c$ for the source currents, only enters into the problem at the interfaces.

It is instructive at this stage to examine the set of equations generated to determine ϕ and $\partial\phi/\partial n$ in a two-region problem, where there is an interface between the regions. A pictorial representation of the equation is shown in Figure 10.9. There are $n1$ and $n2$ sides and $m1$ and $m2$ unknowns in region 1 and 2 respectively. The submatrix (1) is dense and is formed from the coefficients from Equation (10.27b) applied to the element of region 1. Similarly submatrix (4) comes from region 2. The submatrices (2) and (3) are sparse (two unknowns per row) and are generated from the interface conditions. The other areas contain zeros. If on the boundary surfaces where the

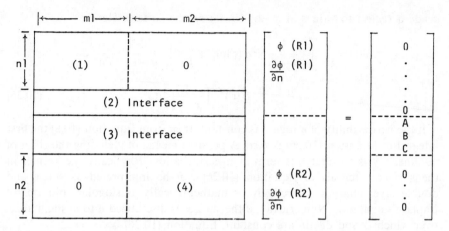

Figure 10.9   A representation of the set of linear equations required to solve for the potential and its normal derivative in a 2-region problem

potential or its derivative is known the value is zero then all the right-hand sides are zero except those corresponding to the normal **B** continuous boundary conditions (*A*, *B*, *C* in Figure 10.9).

10.5.2   Application of the method

A two-dimensional magnetostatic computer program was written to assess the accuracy and efficiency of the method. The boundaries surrounding regions of different permeabilities were subdivided into plane faces over which the potential and its normal derivatives were assumed constant. The integrals in Equation (10.27a) for the matrix coefficients are easily expressible in closed form.[17]

Any symmetry in the problem is utilized to reduce the number of unknowns; for example, in Figure 10.10 the whole magnet is drawn but, because the potentials in the 2nd, 3rd, and 4th quadrants have an exact equivalence to those in the first quadrant, the potentials in the first quadrant are the only ones which must be computed explicitly. The far field boundary shown in Figure 10.10 can be expanded to infinity because there are no boundary connections between it and the magnet; the far field boundary then has no effect on the problem whatsoever. This is obvious for real problems where the potential and its normal derivative to the far boundary can be defined as zero. It is not immediately clear in the two-dimensional infinite limit because the potential from a boundary side becomes infinite at large distances. However, the divergence of the potential from a complete surface must be zero and therefore the contributions from all elements of a surface will cancel to produce zero potential at infinity. The program can be run interactively on the

FAR FIELD ┆ BOUNDARY

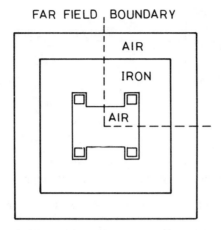

Figure 10.10 Boundary Integral Method model using equivalent elements and symmetry—the far field boundary is shown, but it can be at infinity

Rutherford Laboratory IBM 360/195 and in this version an elegant data input package was used for specifying the boundary data of polyhedra.

10.5.3 Results

The results from two test cases are included in this section: a comparison of analytic and computed results for the field in a hollow, infinitely long, constant permeability cylinder in a uniform external field; and a comparison of the GFUN and Boundary Integral Method computed field for a two-dimensional C-magnet.

10.5.3.1 *Hollow cylinder*

The fields in a hollow, infinitely long, constant permeability cylinder in a uniform field perpendicular to the axis of the cylinder were computed using the Boundary Integral Method. The inside radius of the cylinder was 5 cm and the outside radius 10 cm. The cylinder was approximated by many-sided polyhedra and symmetry was used so that only potentials and derivatives in the first quadrant were computed explicitly. In Figure 10.11 the computed shielding factor of the cylinder is plotted as a function of the number of boundary faces for a cylinder with a relative permeability of 100. The accuracy is very good, and most of the error is due to the polygonal approximation. The field in the hollow centre should be uniform and in the computed cases the homogeneity was always better than 2 in 10^4.

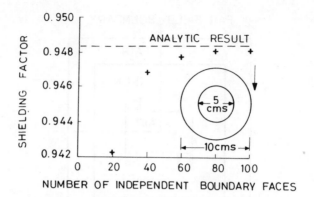

Figure 10.11　Shielding factor of hollow ferromagnetic cylinder—inside radius 5 cm, outside radius 10 cm, permeability 100—as a function of the number of independent boundary faces in the model

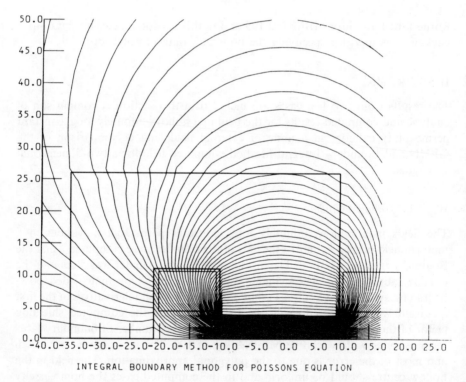

INTEGRAL BOUNDARY METHOD FOR POISSONS EQUATION

Figure 10.12　A map of the computed scalar potential for the C-magnet of Figure 10.1

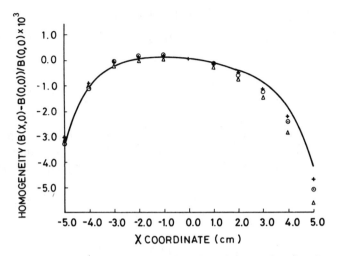

Figure 10.13 Computed homogeneity of the field under the pole tip of the C-magnet of Figure 10.1.——200 element GFUN 3-D, $B_0 = 0.3031$ T (150 s); + 180 element BIM, $B_0 = 0.3035$ T (35 s); ⊙ 140 element BIM, $B_0 = 0.3027$ T (15 s); △ 70 element BIM, $B_0 = 0.3019$ T (5 s)

10.5.3.2 *C-shaped dipole magnet*

The geometry of this magnet is shown in Figure 10.12. The results in Figure 10.1 have shown that GFUN gives accuracies of the order of 0.01% for the homogeneity of this type of C-magnet. GFUN was therefore used to compute the field homogeneity of the magnet shown in Figure 10.12 for steel with a relative permeability of 1000. In Figure 10.13 the GFUN results are compared to those obtained using the Boundary Integral Method (BIM) for several different models. Symmetry was used and therefore only the upper half plane was computed explicitly. (In both cases the far field boundary was at infinity.) The results for this case are again good. Figure 10.12 shows a computed map of lines of constant scalar potential for the 150 element BIM model.

10.6 EDDY CURRENTS IN INTEGRAL EQUATION FORMULATION

10.6.1 Basic equations

The field equations needed for eddy current problems are in SI units.

$$\text{curl } \mathbf{E} = -\frac{\partial \mathbf{B}}{\partial t} \tag{10.28}$$

$$\text{curl } \mathbf{H} = \mathbf{J} \tag{10.29}$$

$$\text{div } \mathbf{B} = 0 \tag{10.30}$$

$$\text{div } \mathbf{J} = 0 \tag{10.31}$$

$$\mathbf{J} = \sigma \mathbf{E} \tag{10.32}$$

$$\mathbf{B} = \mu \mathbf{H}. \tag{10.33}$$

Displacement currents are neglected, free charges can be present only on surfaces and Ohm's law applies. An integral equation is most simply obtained for the potentials $\mathbf{A}$, ψ as follows. Substitute the defining equation $\mathbf{B} = \text{curl } \mathbf{A}$ in (10.28)

$$\text{curl}\left(\mathbf{E} + \frac{\partial \mathbf{A}}{\partial t} \right) = 0 \tag{10.34}$$

$$\mathbf{J} = \sigma \mathbf{E} = -\sigma\left(\frac{\partial \mathbf{A}}{\partial t} + \text{grad}\psi \right) \tag{10.35}$$

for some scalar potential ψ. For a region where the conductivity σ is constant (10.28) implies that ψ is harmonic. If the eddy currents $\mathbf{J}$ are to be calculated in some conducting region under the influence of a known driving field, the solution of (10.29) (in the Coulomb gauge) is:

$$\mathbf{A}(\mathbf{r}) = \mathbf{A}_0(\mathbf{r}) + \frac{\mu_0}{4\pi} \int \frac{\mathbf{J}(\mathbf{r}')}{|\mathbf{r} - \mathbf{r}'|} dV \tag{10.36}$$

in which $\mathbf{A}_0$ represents the driving field and the second term is the vector potential at $\mathbf{r}$ due to the eddy currents. Substituting (10.35) into (10.36) yields the required integral equation for $\partial \mathbf{A}/\partial t$

$$\mathbf{A}(\mathbf{r}) = \mathbf{A}_0(\mathbf{r}) - \frac{\mu_0 \sigma}{4\pi} \int (\partial \mathbf{A}(\mathbf{r}')/\partial t + \text{grad}\psi(\mathbf{r}')) \frac{dV}{|\mathbf{r} - \mathbf{r}'|}. \tag{10.37}$$

The equation for ϕ may be obtained by applying the Coulomb gauge condition ($\text{div } \mathbf{A} = 0$) to this. Alternatively, since ψ is harmonic its value at any point may be expressed by Green's second theorem in terms of the value of ψ and its normal gradient over the surface of the conductor, where the condition that no charge can leave the surface can be written by (10.35) as $(\partial \mathbf{A}/\partial t + \text{grad}\psi) \cdot \mathbf{n} = 0$. Here $\mathbf{n}$ is a unit vector normal to the surface. Hence:

$$-4\pi\psi(\mathbf{r}) = \iint \left\{ \frac{1}{|\mathbf{r} - \mathbf{s}'|} \frac{\partial \mathbf{A}(\mathbf{s}')}{\partial t} + \right.$$

$$\left. + \psi(\mathbf{s}')\text{grad}\left(\frac{1}{|\mathbf{r} - \mathbf{s}'|} \right) \right\} \cdot d\mathbf{s} \tag{10.38}$$

where the source point $\mathbf{s}'$ now varies only over the surface of the conductor.

(10.34) and (10.35) together determine the problem. The components of $\partial \mathbf{A}/\partial t$ are linked only by the surface condition (10.35).

10.6.2 Discretization

Suppose the conductor to be divided into N elements, L facets of these elements forming the surface of the conductor.[3] The basis assumed is then constant $\partial A/\partial t$ and ψ in each element or on each facet, leading to a set of $3N + L$ simultaneous equations in a manner very similar to the $3N$ equations for magnetization in Section 10.3.1

$$
\begin{bmatrix} G_A & G_B \\ G_C & G_D \end{bmatrix} \begin{bmatrix} \partial A/\partial t \\ \psi \end{bmatrix} = \begin{bmatrix} A - A_0 \\ 0 \end{bmatrix}.
\tag{10.39}
$$

Solution of these equations gives values for $\partial A/\partial t$ in terms of the still unknown values of A. The solution is completed by solving these $3N$ first-order differential equations in terms of the initial conditions, e.g. $A = A_0$ at $t = 0$.

Carpenter and Yeh[4] obtain a solution of (10.37)–(10.38) by expressing $\mathbf{A} - \mathbf{A}_0$ as a perturbation series, with the driving field fitted to a polynomial in time. Where this approach converges, it offers considerable economies in the computer space required.

10.6.3 Results

Biddlecombe *et al.*[3] have obtained results with this technique for two-dimensional problems with the driving field along or perpendicular to the axis of no variation. For the case with the current parallel to this axis there is no need for surface charges to prevent a flow of current out of the conductor and the ψ term in (10.37) can be dropped.

Comparisons of the results can be made with analytic solutions for a hollow cylinder with driving field perpendicular to the axis and a rectangular bar with the field along the axis, both driving fields being uniform step functions. A typical result is shown in Figure 10.14. For times short compared to the fundamental mode of the system, the representation converges rapidly.

A more complex geometry is illustrated in Figure 10.15. Here the driving current in the outer bars was sinusoidal for $t > 0$ and the field contours show the response of the rectangular tube at the 1st zero of the 50 Hz drive. Since the current flow is perpendicular to the paper no difficulties result from lapping the elements at the corners.

Current flow in the plane of the paper is illustrated by Figure (10.16). For all problems in this class the eddy currents flow in a solenoid pattern so that they have no field outside the conductor. Thus the 1 Tesla contour should follow the conductor boundary at all times.

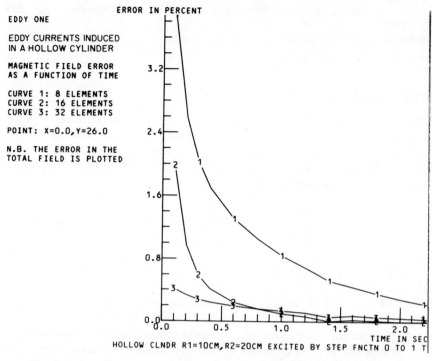

Figure 10.14 Comparison with the analytic result for a step function field

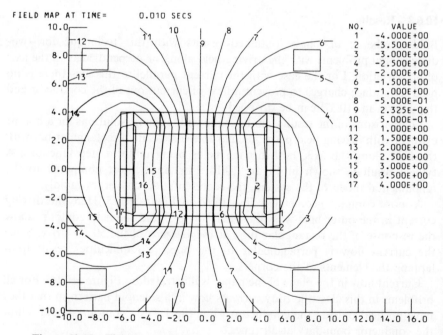

Figure 10.15 Field contours for a copper box driven by two pairs of bars

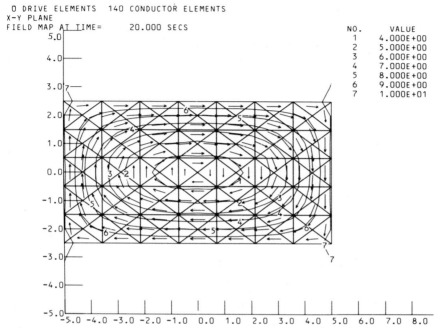

O DRIVE ELEMENTS   140 CONDUCTOR ELEMENTS
X-Y PLANE
FIELD MAP AT TIME=   20.000 SECS

NO.	VALUE
1	4.000E+00
2	5.000E+00
3	6.000F+00
4	7.000E+00
5	8.000E+00
6	9.000E+00
7	1.000E+01

Figure 10.16   Field contours and current flow pattern for a rectangular bar with the uniform driving field along its axis

10.7   CONCLUSIONS

It has been shown that the magnetization integral equation method applied to magnetostatics problems is capable of high accuracy in two dimensions and modest accuracies in three dimensions. In order to reduce the computer resource requirements, especially onerous for complex geometries, two alternative methods have been investigated. Both of these are still at an early stage of development but show, theoretically at least, a clear advantage over the method used in GFUN. For example, Table 10.2 compares projected comput-

Table 10.2

Volume elements	Surface elements in BIM	GFUN Magnetization method		BIM		Scalar int. eq. method	
		Int. eq.	Single field point	Int. eq.	Single field point	Int. eq.	Single field point
216	216	114	2.16	12	0.36	4.2	2.16
343	294	450	3.43	30	0.49	18.0	3.43
512	384	1488	5.12	70	0.64	55.0	5.12
730	486	4320	7.30	133	0.80	162.0.	7.30

ing times (seconds) for a range of elements, i.e. for existing GFUN, Boundary Integral Method and the scalar potential integral equation. The table also gives times for computing a single field point.

As regards the application of these methods to eddy-current problems, it has been established that the integral equation formulation leads to a useful computer program for the two-dimensional limiting cases. The exploitation of symmetry would considerably increase the detail in which symmetrical geometries could be mapped. Improved accuracies can be hoped for by using a higher order basis for the variation within elements, and also by using a more suitably distributed element mesh within the conductors. The coupling coefficients can be evaluated in a general three-dimensional element.[6] Collectively this should result in a general three-dimensional program for iron-free problems.

The major task is to bring the induced current and magnetization formulations of the Integral Equation Method together and to compare the result with the Finite Element Method, or with a Boundary Integral Method. Any very general program along these lines may be prohibitively expensive and there is clearly a place for steady state versions, or, for linear problems, a Fourier transform version.

REFERENCES

1. Armstrong, A. G. A. M., Collie, C. J., Diserens, N. J., Newman, M. J., Simkin., J., and Trowbridge, C. W. (1975). 'New Developments in the Magnet Design Program GFUN', RL-75-066, Rutherford Laboratory.
2. Armstrong, A. G. A. M., Simkin, J., and Trowbridge, C. W. (1975). 'Shims for the EPIC Prototype Dipole Magnet (Calculations)', RL-75-013, Rutherford Laboratory.
3. Biddlecombe, C. S., Collie, C. J., Simkin, J., and Trowbridge, C. W. (1976). 'The Integral Equation Method applied to Eddy Currents', RL-76-043, Rutherford Laboratory. Also *Proc. COMPUMAG Conf, Oxford*, 1976.
4. Carpenter, K. H. and Yeh, H. T. (1976). 'Perturbation Expansion with Separated Time Dependence for Eddy Current Calculations', *Proc. COMPUMAG Conf., Oxford*, 1976.
5. Collie, C. J., Diserens, N. J., Newman, M. J., and Trowbridge, C. W. (1973). 'Progress in the Development of an Interactive Computer Program for Magnetic Field Design and Analysis in Two and Three Dimensions', RL-73-077, Rutherford Laboratory.
6. Collie, C. J., (1976). 'Magnetic Fields and Potentials of Linearly Varying Current or Magnetisation in a Plane Bounded Region', RL-76-037, Rutherford Laboratory. Also *Proc. COMPUMAG Conf., Oxford*, 1976.
7. Cruse, T. A. (1973). 'Application of Boundary Integral Equation Method to 3D Stress Analysis', *J. Computers and Structures*, **3**, 509–527.
8. Grillet, J. P. (1975). Private communication.
9. Harrington, R. F. (1968): *Field Computation by Moment Methods*, McMillan, London. p. 25.

10. Iselin, Ch. (1976). 'A Scalar Integral Equation for Magnetostatic Fields,' *Proc. COMPUMAG Conf., Oxford*, 1976.
11. Jawson, M. A. (1963). 'Integral Equation Methods in Potential Theory 1', *Proc. Roy. Soc., A*, **275**, 23–32.
12. Lachat, J. C. and Watson, J. O. (1975). *Boundary Integral Equation Method: Computational Applications in Applied Mechanics*, The American Society of Mechanical Engineers.
13. Newman, M. J., Trowbridge, C. W., and Turner, L. R. (1972). *Proc. 4th Int. Conf. on Magnet Technology, Brookhaven.*
14. Newman, M. J. (1976). 'A Fortran Package for Solving Linear Algebraic Equations with a Large Dense Matrix using Direct Access Disk Storage', RL-76-018, Rutherford Laboratory. Also *Proc. COMPUMAG Conf., Oxford*, 1976.
15. Parsons, A. (1975). Private communication.
16. *Proceedings of COMPUMAG Conference, Oxford*, (1976), Rutherford Laboratory.
17. Simkin, J. and Trowbridge, C. W. (1976). 'Magnetostatic Fields Computed using an Integral Equation derived from Green's Theorems', RL-76-041, Rutherford Laboratory.
18. Smythe, W. R. (1968). *Static and Dynamic Electricity*, McGraw Hill, New York. p. 53.
19. Symm, G. T. (1974). *Potential Problems in Three Dimensions. Numerical Solution of Integral Equations*, Clarendon Press, Oxford. Ch. 24.
20. Trowbridge, C. W. (1972). 'Progress in Magnet Design by Computer', *Proc. 4th Int. Conf. on Magnet Technology, Brookhaven.*
21. Trowbridge, C. W. (1974). 'Computer Aided Design at the Rutherford Laboratory', RL-74-133, Rutherford Laboratory.
22. Turner, L. R. (1973). 'Direct Calculation of Magnetic Fields in the Presence of Iron, as applied to the Computer Program GFUN', RL-73-102, Rutherford Laboratory.
23. Simkin, J. and Trowbridge, C. W. (1979). 'On the use of the total scalar potential in the numerical solution of field problems in electromagnetics', *International Journal for Numerical Methods in Engineering*, **14**, 423–440.
24. Armstrong, A. G., Collie, C. J., Simkin, J., and Trowbridge, C. W. (1978). 'The solution of 3D magnetostatic problems using scalar potentials', *Proc. COMPUMAG Conf. Grenoble*, 1978.

Index